Communications Technology Guide for Business

For a complete listing of the *Artech House Telecommunications Library*, turn to the back of this book.

Communications Technology Guide for Business

Richard Downey
Seán Boland
Philip Walsh

Artech House
Boston • London

Library of Congress Cataloging-in-Publication Data
Downey, Richard.
  Communications technology guide for business / Richard Downey, Seán Boland,
Philip Walsh
    p.   cm. — (Artech House telecommunications library)
  Includes bibliographical references and index.
  ISBN 0-89006-827-5 (alk. paper)
  1. Business—Communication systems.   2. Information technology—Equip-
ment and supplies.   3. Internet (Computer network)   I. Boland, Seán.
II. Walsh, Philip.   III. Title.   IV. Series.
  HF5541.T4D68 1997
  651.8—dc21
                                                    97-32245
                                                    CIP

British Library Cataloguing in Publication Data
Downey, Richard
  Communications technology guide for business
  1. Telecommunication systems   2. Business–Data processing
  I. Title II. Boland, Seán   III. Walsh, Philip
  621.3'82

  ISBN 0-89006-827-5

Cover design by Deborah Dutton and Joseph Sherman Design
Microsoft and Lotus Freelance clip art is used in this book, with permission.

© 1998 ARTECH HOUSE, INC.
685 Canton Street
Norwood, MA 02062

International Standard Book Number: 0-89006-827-5
Library of Congress Catalog Card Number: 97-32245

10 9 8 7 6 5 4 3 2 1

Contents

Preface

In today's telecommunications industry, there is a somewhat bewildering set of technologies and services available. Some of these are destined to change the world we live in, others have enormous significance in niche markets, while others will inevitably be relegated to the status of "old technology" and disappear slowly or rapidly.

Most business managers already sense that they could be making better use of communications technology. They have seen and heard examples of success elsewhere, but they do not feel qualified to make a plan for their own business. Employing a consultant or inhouse expert to make a plan is one option. However, ideally the manager should be taking an active part in the planning process and must at least understand what the experts are proposing.

In business, there is often an inertia towards the adoption of new technologies. Delayed implementation contains an implicit risk of a missed, or at least delayed, opportunity. Many want evidence of concrete success stories and monitor how others get on with a technology. Others wait to allow time for any initial problems to be solved and for costs to fall. However, often the reason for not implementing new technologies is a lack of understanding of their capabilities and the opportunities that they create. In today's world of constant change and continuous improvement, unwarranted delays may well leave a business at a significant disadvantage.

The growth of the Internet is probably the best known example of the way communications technology is creating new opportunities for business. Many have already embraced it wholeheartedly. Others, however, have been slower or less informed in their approach to this technology. If this is the case for such a visible technology, it is not surprising that the situation is much worse for technologies that do not attract such media hype. Video conferencing, for example, has great potential to save time and money and is already a relatively mature technology, but the number of adopters is still quite low.

The situation is even more confused when it comes to distinguishing between such services as X.25, ISDN, and frame relay. Some of the newer services have the potential to significantly reduce communication expenditure in certain instances, but this may go unrecognized for quite some time if some basic knowledge is not available inhouse.

Security is an issue that some tend to overlook in their eagerness to gain some business advantage, while others are paralyzed into doing nothing for fear of major problems. An understanding of security issues will help to avoid either extreme.

Aim of This Book

This book aims to give managers with communication responsibilities and indeed general business managers a broad view of the technologies available today, so they can deal knowledgeably with consultants, suppliers, and inhouse experts and make informed choices. We expect our primary readership to include business managers and financial personnel. With this in mind, we believe that this will be a useful text for many business courses, particularly those that include modules on the application of technology to business. We also envisage that this book will provide a useful introduction/overview for communication managers, IT managers, and telecommunication engineers. Finally, we expect sales and marketing personnel and many others who work in the telecommunications industry to benefit from this book.

The basic principles are explained so that the reader is better able to distinguish between technologies. Typical applications are described and cost issues are discussed. The inevitable limitations—so often missing from sales literature—are also explained. Security issues are discussed so that the reader will know what parts of a communication system are vulnerable to attack and can arrange appropriate protection.

The development of appropriate communication solutions and their subsequent implementation are also discussed. A business-planning process is described, followed by recommendations on developing a solution, procurement, and supplier management. The impact of tariffs on any proposed solution is also addressed. We strongly recommend that readers who are more technically minded do not skip over this part of the book, as it introduces topics that are sadly missing from many communication projects driven primarily by technical considerations.

This book has grown out of a seminar entitled "Business Communications" developed and delivered by the authors in conjunction with Telecom Ireland International, the contracting and consultancy wing of Telecom Eireann. Through these seminars, the authors have gained an insight into the material of most interest to an audience of business managers and communication managers. The opinions expressed in the book are those of the authors and not necessarily those of Telecom Eireann.

Telecom Ireland International specializes in providing training, consultancy, and operations support in the telecommunications industry and can be contacted by sending email to tii@telecom.ie.

Reading Guidelines

This book has been arranged to facilitate the reader who wishes to read it from start to finish. We realize that such a reading plan will not suit many busy readers, and thus we have tried to write individual chapters in such a way that they could be read on their own aided by a few cross-references and glossary lookups. In order to gain the most from this approach we recommend that such a reader would first familiarize themselves with the contents of Chapter 2. Furthermore we suggest that a newcomer to communications technology might also read the first few pages of each chapter in the first part of the book in order to get a better overview of things.

The glossary has been compiled with the "chapter at a time" reading plan very much in mind. We have limited the contents of the glossary to terms which crop up in multiple chapters. Thus, for example, if you were to launch into the book at Chapter 6 on ISDN, you would find ISDN terminology explained within the chapter, and most general telephony jargon that appears unexplained would be found in the glossary. Much of the ISDN-specific terminology is not, however, included in the glossary. If necessary these terms can be found using the index.

The references at the end of the chapters contain many web references. It is the nature of the web that these references will no longer be valid in the future. Where this is the case it may still be possible to find the reference through the use of a general search engine such as AltaVista. Alternatively, you could find the home page of the particular site and browse from there to find the article. You might also check if the particular site has its own search engine. This way you may find an update to the original article. You can find the home page of most sites by typing in the web reference ignoring everything after the domain name (e.g., typing http://www.*domainname*.com instead of, say, http://www.*domainname*.com/bla/bla/bla).

In many of the diagrams we represent networks using a cloud symbol. Newcomers may find this confusing. The cloud is simply used to represent a more or less complex network in the same way as a box might be used to represent a complex piece of equipment. This is quite a standard approach, which some jokingly refer to as *cloud technology*.

Acknowledgments

First, we would like to thank Paul Thornton, LAN administrator, pricing division, Telecom Eireann, who wrote Chapter 12 on LAN interconnection and helped a lot with the material on LANs in Chapter 4. We also appreciate all at Artech House for their patience and perseverence with the project. A special word of thanks to the Artech House technical reviewer for his helpful and pertinent comments and his assistance in giving a U.S. slant on the subjects and to Glenn Powell, vice president of Valuecom for his help in reviewing the chapter on tariffs. We also wish to thank Telecom Eireann for their support.

We are grateful to those who checked our work for relevance and readability, including Luis Lasalla, Barry O'Grady, Daniel Dorn, Carlo Papa, Juan Martinez, James Hurley, John Cahir, Aidan Scanlon, and Pat Feenan. We are also glad to acknowledge the technical input received from the staff of Telecom Eireann, including Sean Abraham, Tim Flynn, Paul Dervan, Denis Curran, Derik Frier, Vincent Hendrick, Liam O'Connor, Pamela Harisson, Pat Duggan, Philip Maguire, Charlie Tone, Neil O'Sullivan, Paul Grealy, and all those who contributed indirectly by sharing their knowledge and experience with us over the years.

Our thanks to Peter Cullen at Trinity College Dublin; Donal McGuinness at Lake Electronics; Chris Coughlan, marketing and strategic planning manager, Digital Equipment International; and Tom Quinn, regional sales manager at Tellabs. Next, we thank the ever-helpful Mary O'Dowd, librarian of the Telecom Eireann corporate library, who helped us so much in our search for relevant material, and Eddie Cahill, who provided much-needed encouragement during the initial development of the "Business Communications" seminar.

Finally, we thank Mary Boland, Antoinette Walsh, and all the members of our families who have borne with us patiently and offered us much needed support and encouragement. Our children, Amy and Ruth Boland and Steven and Brandon Walsh deserve special mention. All will be happy that the project is finished, and Amy and Ruth can, at last, stop asking, "When will daddy be finished with 'the book'?"

1

Communication Services and Networks

1.1 Introduction

Communication services, and their associated communication infrastructure, play a vital role in today's business environment. The list of potential applications is continually growing, and businesses that exploit them to their fullest extent can gain a significant competitive advantage over their rivals. Businesses that do it well are reaping significant benefits when the specified solutions are finally implemented. They are gaining new customers, accessing new markets, and generally thriving—often at the expense of their less communication-aware competitors.

Current and emerging technologies now enable even the smallest businesses to take advantage of a wide range of communication services. This means that, to stay competitive in this environment, many businesses will have to become significantly more "communication aware" than heretofore. They will have to make greater use of communications technology as an everyday business tool and do so in an efficient and innovative manner.

Although this book is primarily about communications technology, the authors are fully aware that a distinct dividing line between communications technology and *information technology* (IT) no longer exists. Throughout the book the term *information and communication technologies* (ICT) is used as a general term when dealing with information electronically. The term *communications technology* is used when we wish to highlight the communication-specific aspect of any solution.

1.2　Function of Communications Technology

The basic function of communications technology is to facilitate the transfer of information between computer systems or end users. Put another way, communication systems provide *connectivity*. Ideally this connectivity should be instantaneous—and universally available regardless of time, distance, or location. It should be transparent to all applications and all types of information, with no regional, national, or international variations. In this utopian world, communication networks would be regarded as just another utility analogous to gas or electricity networks. In practice, many constraints—legal, economic, regulatory, and technological—prohibit such universal connectivity.

1.2.1　Types of Connectivity

Connectivity can be either permanent (as in the case of a fixed communication link) or temporary (such as a phone call).

Permanent connectivity is used where:

- Information must be transferred on a continuous or near-continuous basis (e.g., a video surveillance link);
- Immediate connectivity must be available when required (e.g., a link from a financial services company to an information provider);
- It is more cost-effective than temporary connectivity (e.g., between multiple sites belonging to the same business).

Temporary connectivity is used where:

- Information is only transferred on an occasional or intermittent basis (e.g., between a retailer and its suppliers);
- Permanent connectivity is too expensive (e.g., it cannot be afforded or cannot be justified);
- It is a particular requirement of the application (e.g., permanent connectivity cannot be provided between all users of the public telephone network).

1.2.2　Cost of Connectivity

Communication services are a significant cost to any business. The good news is that these costs are still falling in most countries. It is also becoming easier to measure communication costs more accurately. This allows costs be apportioned according to usage or particular requirements specified. In some businesses it also facilitates passing on communication charges to clients. However, it is likely that cost constraints will continue to limit how businesses use communications technology, at least for the moment. Common compromises include restricting the amount of information transferred or delaying the transfer of information until reduced rate tariffs apply.

Connectivity costs are related to the following:

- *Quantity of information:* the greater the quantity of information the greater the cost;
- *Distances involved:* the cost of connectivity increases with distance;
- *Mobility requirements:* the cost of connectivity is significantly greater for mobile applications;
- *Time sensitivity:* connectivity is less expensive during off-peak periods;
- *Duration:* temporary connections are cheaper than permanent connections if the connectivity is only required for short periods.

1.2.3 Scope of Connectivity

Connectivity may be required within a confined geographic area (e.g., a single site) or may be required over a geographically dispersed area. Means of providing single site connectivity include *private branch exchanges* (PBXs) for voice and *local area networks* (LANs) for multiple types of information. These options are covered in Chapters 3 and 4, respectively. Single-site communication solutions can be completely controlled by a business and, optionally, owned by it. In addition, due to the short distances involved, high-performance networks can often be installed cost effectively.

Numerous options are available for intersite connectivity and many of these are discussed in Chapters 5 through 9. Where temporary connectivity is required, a number of public networks (such as the public telephone network) are available. These are often referred to as *wide area networks* (WANs). They can be further characterized as being either national or international. Where connectivity is required on a frequent or continuous basis a private WAN can also be used.

Private WANs are expensive to install and operate, more particularly due to the high cost of the intersite communication links. This means that many businesses cannot afford them. Fewer still can afford to run information-intensive applications, such as multimedia, at their full potential over WANs. International WANs represent an even greater challenge as multiple suppliers of equipment and communication links may be required. *Service integrators* are often used for such networks because of their knowledge as to what can and cannot be done in the different countries.

1.3 Communications Technology and Business

From a business perspective, the overriding concern is to identify how best to deploy communications technology to gain maximum competitive advantage. This means having a good appreciation of the many communication services now available and how they might be used. Table 1.1 illustrates how communications technology, often allied with other aspects of information

Table 1.1
Typical Business Applications

Business Activity	Communication/IT Services
Advertising	World Wide Web (WWW), email, and fax with client's agreement
Sales	WWW, telesales (call centers)
Sales support	Online information services, multimedia
Customer profiling	Networked client-server databases
Ordering/invoicing	Electronic data interchange (EDI)
Payment	Electronic funds transfer/electronic cash
Customer support	Help desks, email, WWW, fax, bulletin boards, online information services
Purchasing	Public online information services, WWW
Meetings/discussions	Voice and data calls, video telephony/conferencing
Messaging	Voice mail, email, data pagers, wireless applications
Answering routine inquiries	Interactive voice response (IVR) systems, fax, WWW
Project work	Voice and data calls, fax, email, groupware, intranet
Recruitment	WWW, newsgroups
Document storage and retrieval	Combined document image and electronic document database, intranet
Document distribution	File transfer, email, fax, WWW, intranet
Software distribution	File transfer, WWW, bulletin boards
Research and development	Online information services, WWW, virtual conferences, special-interest groups

technology, can assist with a wide range of business activities. The list is meant not to be exhaustive, but to illustrate the extensive range of possibilities currently available. Information on the various services can be found later in this book.

1.3.1 Some Communication-Enabled Companies

This section illustrates how ICT can be used by the different business sectors. A number of examples, covering all business sizes, are given. There is a perception that only large businesses can benefit from the more sophisticated applications of communications technology. This is not true! It is certainly the case that larger businesses more readily adopt sophisticated communications technology applications. However, many applications, such as computer telephony integration, fax, the WWW, and voice mail, to mention but a few, are within the reach of even the smallest business.

Retail

Virtual-shopping is becoming an ever-more feasible option with a wide range of goods and services available online. This trend has been fueled by the explosion in the web, where everything from books to luxury food items are now offered for sale. An IT center in Donegal, Ireland [1], has helped a range of businesses take advantage of this sales avenue. It develops multimedia catalogs for clients, both large and small, which can then be made available online. Similar services are available from many other companies.

Mass customization, where individual preferences can be incorporated into products or services, is often critically dependent on ICT. Examples of this growing phenomenon can be found across all sectors ranging from jeans manufacturers to suppliers of agricultural machinery [2]. A chain of luxury hotels uses ICT to compile a database of individual customers' preferences, while a grocery retailer now offers customized orders and customized loyalty benefits [3].

A pizza company uses *computer telephony integration* (CTI) to help build a database of customers. When new customers phone in orders, their telephone numbers and other details are logged. On repeat calls, this information is automatically presented on a computer screen, allowing more efficient and more customer-friendly service. The frequency and value of sales to each customer are also tracked automatically. This facilitates ongoing marketing campaigns.

Travel and Tourism

A large American-based hotel chain has its own private, worldwide, communication network [4]. In addition to supporting common business functions, such as reservations and checking in and out, this network also supports an extensive data warehousing facility. All transactions, from anywhere in the world, reach a central database within a number of hours. The information held in this database can then be used by business analysts to set seasonal pricing levels, for example.

The Barry House Hotel is a small London hotel near Hyde Park with just four employees. By advertising on the Internet, they gained an additional 50 inquiries a week [5].

Marketing

A private WAN has helped an international advertising firm manage clients' accounts more effectively while reducing cost [6]. The network links teams in diverse geographic locations and carries advertising copy and artwork as well as voice. A global headhunting firm also uses a private WAN to link consultants in numerous small offices [7]. Each consultant can access the company database and access or update information as required.

Advanced CTI facilities have transformed a New York–based market research firm. The system allows auto dialing from a computer database, and research interviews are digitally stored for future analysis [8].

Wholesale and Distribution

A British container-handling facility [9] uses a mixture of fixed and mobile technologies to provide a range of online information at all points in the logistics chain. Using EDI, information is not only provided internally but is also passed electronically to outside agencies, such as shipping lines and customs and excise organizations. EDI also plays a central role in an American coolant firm's marketing and distribution strategy. It has helped them capture greater market share while increasing efficiency through processing a greater proportion of sales electronically [9].

Satellite technology has helped an Irish freight-forwarding company gain a significant competitive advantage [10]. It uses mobile data terminals and the *global positioning system* (GPS) to allow automatic vehicle location and consignment tracking. This facilitates greatly improved customer service, while invoicing can now proceed as soon as a consignment is delivered.

Manufacturing

Information and communication technologies are helping manufacturers reduce costs, improve efficiency, and give better customer service. EDI facilitates just-in-time operation at all points in the delivery chain. Goods can now be ordered, manufactured, and delivered on a near-demand basis with consequent saving in inventory costs. Geographically dispersed design teams can be linked by videoconferencing links and *whiteboard* software for collaborative working.

The process can be extended further by means of high-capacity WANs that allow team members work together on computer-aided design files in real time.

Financial Services

Nowhere is ICT more widely deployed than in the financial services sector. Many of the facilities we now take for granted (e.g., the ubiquitous automatic teller machine) depend heavily on the use of communications technology. Online banking, direct sales—the list is endless. EDI is opening up new opportunities and is now used by many financial services companies to reduce back-office costs and allow more resources be deployed for serving clients.

More specialized communications applications are also used by the financial services sector. A German bank launched an automated telephone banking system using voice-recognition equipment [11]. Its ability to operate without the need for touch-tone phones was a primary issue, as much of the target population had rotary telephones. Meanwhile, automatic identification of the calling number (A-number) is a key business tool for a U.K.-based direct sales company whose portfolio includes mortgages and life insurance [12]. Existing clients, whose details are automatically loaded onscreen, are always directed to the same agent if available. Any unprocessed calls are given a higher queue priority on any subsequent calls. Finally, repeat queries, from the same telephone number, are readily identifiable to avoid skewing any statistical analysis.

Education and Training

The potential for using ICT in the education and training sector is immense. Learning can be personalized, interactive, and explorative rather than passive and text based. Techniques such as *computer-based training* (CBT) and *computer-aided training* have been widely used for some time, as have *compact disc–read only memory* (CD-ROM) materials. What is changing is the level of distance-independence now possible with modern communications technology and the ease of access to the latest information from around the world.

The range of information now available online is truly staggering. Material from the U.S. Library of Congress, the National Library of Canada, and many other libraries, museums, and national archives is being digitized and put online. Many other bodies, both voluntary and commercial, also make information available. Even more impressive than the volume is the sheer diversity. Photos, illustrations, audio, video, and animation are all available and often far more contemporary than could ever be found in any traditional library or course notes.

An American business school, in collaboration with local universities and other sponsors, offers an MBA to a number of Pacific Rim countries via distance learning. It uses satellite-based, two-way, full-motion videoconferencing, allied to data communication and Internet-based course materials [13]. Another American university has built a *virtual college,* where students can attend classes anytime anywhere. The technology used includes email, digital video, and collaborative software.

Hospitals and Health Care

Communication is playing an ever-increasing role in the medical and general healthcare areas. A New England medical center [14] uses electronic image transfer to allow specialists in Boston to view fetal ultrasound images from clinics across Massachusetts. Meanwhile, a telemedicine network in Marquette, Michigan conducts remote psychiatry sessions by videophone [14]. As medical technology develops, the quantities of information to be transferred in telemedicine applications continue to increase. As part of this trend, a number of Irish hospitals have completed successful trials of high-capacity public networks that allow advanced diagnostic procedures to be carried out remotely.

A large hospital in Denver [15] demonstrates another approach to harnessing the power of communication. It uses wireless laptop computers to ensure the ready availability of clinical information on all patients. The system also makes the hospital's knowledge base available to its clinicians. Not all medical applications involve diagnosis or general management—communication systems also have a role to play in improving the patients' overall environment. One particular project [16] combines computer networks and videoconferencing to enable sick children to communicate and play games from their hospital beds.

1.4　Developing Communication Solutions

As we have just shown, the range of applications for communications technology is extensive. What the examples given don't illustrate is the level of work, and expertise, necessary to develop many solutions. The authors believe that having a basic understanding of the underlying technology is of great benefit to individuals involved in this area. This is true regardless of whether the final solution is developed solely in-house, with the help of a consultant or fully outsourced. Such an understanding is also helpful for managers who, while not directly involved in development work, have to give financial approval or other backing for a project.

In the following chapters, we first introduce the main technologies used in modern communications solutions. The treatment, while not exhaustive, is intended to give a good grounding in this area. It should leave readers well equipped to carry out any additional research that their particular circumstances may require. Next, the development of a communications solution is considered from a business perspective. In particular, communication-specific issues, which may not be readily apparent to the nonspecialist, are highlighted.

References

[1] Information Technology Centre, Letterkenny, Donegal, Ireland.

[2] Moad, Jeff, "Let Customers Have it Their Way," *Datamation,* April 1, 1995.

[3] Laberis, Bill, "Ten trends that will shape your network," *Network World,* April 29, 1996.

[4] Liebman, Lenny, "Keeping Inn Control," *LAN Magazine,* June 1995.

[5] Information Society Initiative, 1996 U.K. Government Web Site, http:/www.isi.gov.uk/

[6] Mendler, Camille, "How To Get Ahead In Advertising," *Personal Computer World,* May 1995.

[7] Liebman, Lenny, "Casting a World-wide Net," *LAN Magazine,* January 1996.

[8] "Market Research Firm With Moribund Power Dialler Rescued by CTI," *CTI for Management/Telemarketing,* December 1995.

[9] "Letting EDI off the Leash," *Communication Networks,* January 1995.

[10] Franzon, Göran, "Mobile Data Rolls Out," *Telecommunications™,* March 1995.

[11] Candy, Louise, "Juggling Calls," *Communication Networks,* June 1995.

[12] "CTI Exploits New Sales Channel for Life Insurance," *CTI for Management/Telemarketing,* December 1995.

[13] Feller, Gordon, "The Cyberspace MBA," *Global Telephony,* April 1995.

[14] Weinberg, Neal, "Physicians Use ISDN as Lifeline to Patients," *Computerworld,* March 4, 1996.

[15] Clarke, Elizabeth, "The Wireless Ward," *LAN Magazine,* December 1995.

[16] Ouellette, Tim, "StarBright Net Gives Sick Kids On-Line Relief," *Computerworld,* November 20, 1995.

2

Communication Principles

2.1 Introduction

This chapter introduces some of the basic principles and jargon that will be used in the remaining chapters. Much of the material is relevant to both single-site and multisite networks; however, the emphasis is on multisite networks.

2.2 Analog and Digital

There are two fundamental modes of transferring information: analog and digital. In general, networks that interconnect computers use digital transmission, while some voice networks may use analog transmission over part of the network.

In digital transmission, all information is transmitted as a sequence of 1s and 0s. Using 1s and 0s in this manner is referred to as *binary,* and each 1 or 0 is called a binary digit, or *bit.* The bits are typically transmitted as electrical pulses (Figure 2.1) or, in the case of optic fibers, as pulses of light. Standardized techniques are used to convert voice, text, and images into digital format and are described later.

In the case of analog transmission (which is principally used for voice communication only), the electrical signals mimic the sound waves as shown in Figure 2.2.

The most common example of an analog communication device is the normal household telephone (including those with a touch-tone keypad). The link between this telephone and the *public switched telephone network* (PSTN)

11

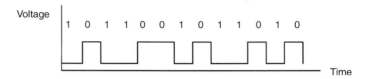

Figure 2.1 A digital signal.

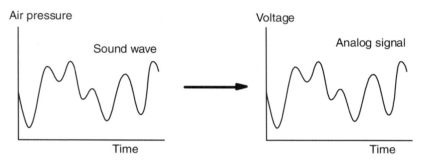

Figure 2.2 Analog transmission: sound waves are converted to an analogous electrical signal.

may also be analog. In most developed countries today, however, the rest of the telephone network (the exchanges and the long-distance links) is digital.

2.2.1 Bandwidth

Bandwidth in a communication context means the information-carrying capacity of a communication link. On a digital link, this is measured in *bits per second* (bps). On an analog link, bandwidth is measured in terms of the range of frequencies that the link can carry measured in *hertz* (Hz). Thus, the telephone network is said to have a bandwidth of 3.1 kHz (1 kHz = 1,000 Hz). This second usage is the original meaning of *bandwidth;* however, in modern digital communication, bandwidth invariably refers to bits per second.

Bandwidth is often quantified in terms of *thousands of bits per second* (Kbps) and *millions of bits per second* (Mbps). The term *speed* is sometimes used instead of bandwidth (e.g., a communication link is said to have a *speed* of 64 Kbps).

2.2.2 Conversion of Information Into Digital Format

One of the advantages of digital transmission is that it can be used to carry all types of information (e.g., text, still images, voice, and video). We will now look briefly at how information is converted into digital format.

Text

Text is usually converted using the *American Standard Code for Information Interchange* (ASCII). This is a convention whereby each letter of the alphabet is represented by a seven-bit binary code. For example: a = 110 0001, A = 100 0001, % = 010 0101. ASCII also contains codes for numerals and the more commonly printed symbols, such as !,:?$*&. In many cases, eight bits are used instead of seven. This allows for accented characters and other symbols. A typical page of text contains about 2,000 characters and can be coded with about 16,000 bits (or 2,000 bytes).

Speech (Audio)

The oldest and the most widespread technique for converting speech into digital format is called *pulse code modulation* (PCM). The signal is first filtered to remove high-frequency components. The voltage is then measured at regular intervals, and each measurement is converted into an eight-bit binary number (Figure 2.3). This conversion process involves an approximation, and, because of this, there will be some encoding noise in the recovered signal. This is normally not objectionable.

The PCM process used in telephone networks produces a 64-Kbps digital signal. This bit rate is very widely used, and it and its multiples have become the standard bit rate for many digital services—including voice. There are two versions of PCM: A-law and mu-law (μ-law). Both techniques are similar, and both use the same 64-Kbps bit rate. The mu-law is used in North America, while the A-law is used in Europe. International circuits linking networks that use different versions must use equipment to perform "A-law to mu-law conversion."

PCM is inefficient by today's standards. Modern voice coding techniques can give similar speech quality using lower bit rates. Table 2.1 gives some examples where low bit-rate speech coding is used.

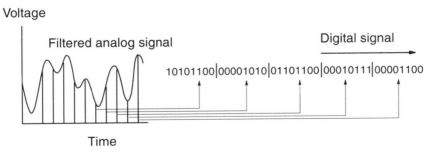

Figure 2.3 The PCM process.

Table 2.1
Low Bit-Rate Speech Coding

Application	Bandwidth	Main Constraint
Digital cordless phones, long-distance links in public networks (ADPCM)	32 Kbps	Quality
Mobile phones, video phones, Internet telephony, and private voice networks	13–6 Kbps	Quality and bandwidth
Satellite communication	4.8–2.4 Kbps	Bandwidth

Low bit-rate speech coding is often used in private voice networks to save on the cost of communication links. Care must be taken to ensure that speech quality is acceptable for any particular application. For a given coding method, lower bit rates imply lower voice quality. There are, however, many examples of new coding techniques delivering *better* speech quality with *lower* bit rates than the older techniques can deliver.

Still Images

Still images are converted into digital format as *bitmaps*. The image is divided up into an array of small points called *pixels*. The number of pixels per square inch determines the *resolution*. Each pixel is coded with a binary number to represent its color and brightness. The more bits in this number, the greater the *color depth*. At the lowest end of the scale, a black-and-white image can be coded with a single bit for each pixel. At the other end of the scale, a color photograph for a publication will probably be coded with 24 bits per pixel.

It is generally possible to considerably reduce the number of bits required to represent an image in digital format using *compression*. The compression can be *lossless,* which does not modify the bitmap image in any way, or *lossy,* which does alter the quality of the image. The designers of telemedicine imaging systems, for example, might worry about lossy compression resulting in mis-diagnoses. In other applications, however, lossy compression is a very worth-while compromise between cost and quality.

Video

The conversion of video into digital format involves the conversion of each video frame into a bitmap. This is most demanding in terms of the number of bits required for a second of video. Even with compression, about 5 million bits are required for a second of broadcast-quality video. With video telephony, on the other hand, clever compromises are made to reduce this requirement to as low as 30,000 bps.

2.2.3 Bandwidth Requirements

The bandwidth requirements of various applications are listed in Table 2.2. This table can be used to gain a basic insight into communication requirements. Some applications require a high bandwidth and, hence, can involve significant costs. The figures given are typical, and the bandwidth requirements can often be reduced at the expense of overall performance.

2.2.4 Digital and Analog Compared

Table 2.3 provides a comparison between analog and digital communication. It should be noted that pure analog networks are becoming very rare. Most of the world's telephone networks are only analog at the extremities—that is, in

Table 2.2
Typical Bandwidth Requirements

Application	Examples	Bandwidth (Kbps)	Special Characteristics
Speech	Standard telephony	8–64	Sensitive to small delays
Still image	Home shopping catalogs	64 +	Depends on image quality
Still image	Medical images	64–1,544	High bandwidth essential in emergency cases
Moving image	Video telephony	30–128	Real time*
Moving image	Video conferencing	128–384	Real time*
Moving image	Broadcast-quality video	5,000 +	Real time* and very high quality
Moving image	Home-quality video (VHS)	1,544 +	Real time*
Transactional	Automatic teller machines	<1	Real time*
Messaging	Email	1 +	Not real time*
Messaging	Electronic funds transfer	1 +	Real time*
Document transfer	Fax	2–64	Not real time*

Real time indicates information that must be transmitted without delay.

Table 2.3
Comparison Between Digital and Analog

Characteristic	Digital	Analog
Application independence	Digital transmission is very versatile and can carry information from a wide range of different applications, including those transferring voice, images, and text.	Analog transmission is optimized for speech.
Cost	Digital communication is often more cost effective for large businesses with a requirement for high-volume voice and high-speed data applications.	Analog speech networks can be very cost effective for the smaller business. Line rental tends to be cheaper than for their digital counterparts as does the associated equipment costs (e.g., phones and fax machines).
Suitability to IT	Much business information to be carried over communication networks is already in digital format (e.g., text documents, computerized databases). Digital networks are ideal for the transport of this type of information.	With analog networks, the digital information must be transmitted using modems with a maximum transmission speed of about 30 Kbps.
Quality	Digital networks offer consistent quality because digital signals suffer little or no degradation within the network (with the exception of mobile radio networks). Any degradation that does occur can be compensated for by using error-correction techniques.	Analog (voice) signals are vulnerable to a wide range of impairments, especially noise at different stages in the communication network. The quality of analog connections can vary greatly and tends to reduce with distance.
Support for advanced features	Communication networks often offer many additional features in addition to basic connectivity. Many of these could not be implemented using non-digital technology.	
Security	Digital signals are more difficult to intercept, as specialized equipment is required. In addition, powerful encryption techniques can be used to give additional protection.	Encryption of analog signals is possible but expensive.
Spectral efficiency	There are only a limited range of frequencies available for radio communication. Digital radio networks can use the available frequencies more efficiently than can nondigital ones.	

the telephone and its link to the nearest exchange. The main links between the exchanges and the exchanges themselves are mostly digital. Analog to digital conversion (PCM) takes place at the input to the exchange.

2.3 Networks

The function of a communication network is to transport information from one device to another. It is normal to categorize networks into *single site* and *multisite* because the equipment, links, and economics of each type can be quite different. Multisite networks are often referred to as *wide area networks* (WANs). Single-site networks are also called *local area networks* (LANs) when computers (as opposed to telephones) are involved.

2.3.1 Communication Links

Communication links provide connectivity between different pieces of equipment. A communication *medium* (e.g., copper cable) is required to actually carry the information signals. Within a single site it is relatively easy to select and install an appropriate medium. In the case of multisite networks, it is rarely practical to install a dedicated medium between each location. Instead, a communication link, with specific information-handling capabilities, is rented from a service provider.

Intersite Links

The vast majority of intersite communication links, whether permanent or temporary, use direct *point-to-point* connections between two locations. An alternative type of link, known as *multipoint*, consists of a single branched circuit and can be used in place of several point-to-point circuits in specialist applications such as automatic teller machines (Figure 2.4). This can be very cost effective where a large number of locations have to be connected. It is not generally suitable for two-way voice communication.

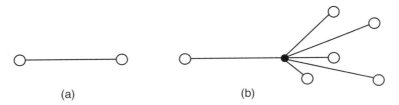

(a) (b)

Figure 2.4 (a) Point-to-point and (b) multipoint circuits.

2.3.2 Communication Media

These are the physical means used to carry the information signal between two locations. A wide range of media are available, each with their own particular characteristics and information-carrying capabilities. They can be divided into three distinct categories:

- *Metallic conductors:* Here, electrical conductors (normally copper) carry the information as an electric voltage or current. Two types of metallic media are in widespread use: twisted-pair cables and coaxial cables (or coax) (Figure 2.5).

- *Optic fiber:* Optical fibers are made from very fine glass fibers encased in a protective material (Figure 2.5). The information is sent through the cable in the form of light pulses. Two types of cable are available: *monomode* and *multimode.*

- *Radio:* Radio is used for a wide range of communication links. It uses electromagnetic waves to carry information and hence is the only category of media that does not require a physical cable. A wide range of frequencies are currently used, ranging from about 20 kHz to 3,000 MHz (1 MHz = 1,000 kHz).

Twisted-Pair Cable

Twisted-pair cable consists of insulated copper wires twisted in pairs to minimize interference. Depending on the application, many pairs can be combined to form a multipair cable. The main limitation of twisted-pair cable is its information-carrying capabilities, which are strongly distance dependent. Typical bandwidths range from 100 Mbps at 100m to 2 Mbps at 4 km.

Twisted-pair cable was the original medium used for telephone communication and is likely to remain in widespread use for some considerable time. It is very flexible in that it will accept any topology, is low cost, and does not take up much room. As a result, it continues to be widely deployed for voice

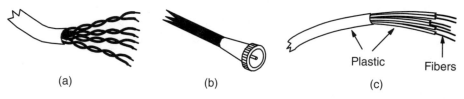

Figure 2.5 Cable types: (a) twisted-pair cable, (b) coaxial cable, and (c) optic-fiber cable.

telephony. It is also in widespread use on LANs as a cost-effective alternative to optic-fiber cable.

Considerable progress has been made in developing techniques for extending the bandwidth of the twisted-pair cable used in the telephone network. The most notable of these are *high bit-rate digital subscriber line* (HDSL), which is used to carry 2 or 4 Mbps over distances of around 4 km using two or three pairs, and *asymmetric digital subscriber line* (ADSL), which carries up to 9 Mbps in one direction over a single pair and about 600 Kbps in the reverse direction [1].

Twisted-pair cable is not a highly secure medium. It is easily "tapped," either physically or by monitoring its electromagnetic radiation. In addition, it is highly susceptible to interference. To overcome these limitations *shielded twisted-pair* (STP) cable can be used. The shielding reduces interference and allows a higher bandwidth at a particular distance. However, it also increases the size and cost of the cable.

Coaxial Cable

Coaxial (coax) cable consists of a central conductor surrounded by a cylinder of fine copper wire mesh and/or an extruded aluminum sleeve. The outer conductor acts as a shield, and coax, hence, has a high immunity to electrical interference. Coaxial cables can support a greater bandwidth than can twisted-pair cable, and bandwidth is not so strongly distance dependent. A typical figure would be 140 Mbps over a distance of 2 km.

This medium is more expensive than either twisted-pair (over short distances) or optic-fiber cable (over longer distances). However, it is far easier and currently less expensive to make multiple connections to coaxial cable than it is to make them to optic-fiber cable. Because of this, it is often used where a number of connections must be made to the same cable. Typical examples include LANs and cable television networks. A very familiar example of coaxial cable is the cable that connects an antenna into the back of a television set.

Optic-Fiber Cable

Optical fibers consist of very fine glass fibers enclosed in a tough outer sheath. Each fiber consists of a central cylindrical core encased by an outer cladding with a slightly higher refractive index. The information is carried over the cable in the form of pulses of light. Light entering the core is continuously reflected at the boundary between the core and the cladding and thus travels along the length of the fiber.

Optical fibers are widely employed to carry telephony and other traffic between telephone exchanges. They are also used for high-speed, high-capacity backbone links on computer networks in large buildings. In this instance, the workstations are still connected via twisted-pair or coax cable and are still subject to the bandwidth limitations of these media.

Optic fibers have a number of advantages over metallic conductors. The biggest one is high bandwidth. A typical fiber allows information to be transmitted at speeds of hundreds of megabits per second over distances ranging from 1 to 200 km. In fact, optic fibers are often viewed as having infinite bandwidth, as their signal-carrying capacity exceeds the capacity of modern electronics to generate the signals!

Because light, and not electricity, is used to carry the signals, optic fibers are not affected by electrical noise. This means that very few errors occur on the transmitted information. In addition, optic fibers are not affected by low insulation and short circuits due to ingress of moisture in the cable (although moisture can adversely affect the mechanical strength of some optic fibers, eventually leading to total failure). The absence of electricity also means that optic fibers can be used in hazardous environments (e.g., with explosive vapors present) where other media would be unsuitable.

Optic-fiber cables are an inherently secure transmission medium. They do not emit electromagnetic radiation and hence eavesdropping is extremely difficult. Even if such an attempt were to be successful, it is possible to detect its occurrence. This is because there will be a detectable drop in signal strength at the receiving end.

Monomode and multimode are the two types of optic-fiber cables available. Monomode cables are superior, supporting greater bandwidth over a longer distances than multimode. Monomode fibers are the favored option for links greater than 6 to 10 km. The cost of both types is similar, but the connectors and the light sources for monomode cables are more expensive.

Radio

The characteristics of this medium vary greatly depending on the particular frequency of the electromagnetic wave. Because of this, it is normal to specify the particular frequencies being used. However, a very high proportion of point-to-point links use frequencies above 100 MHz, and this frequency range is assumed here. Such links can be broadly classified as either fixed or mobile. Mobile communication merits separate treatment and is covered in Chapter 9.

Fixed radio links can be further divided into terrestrial or satellite, although they both share a number of characteristics. Both are used to provide permanent fixed links between individual sites. They are also used to provide temporary and long-distance connections either for a specific event or if a physical cable should fail. Fixed radio links are particularly suited to this latter application, as they can be installed quickly. Many of their advantages arise because no cable is required. In particular:

- They are very flexible and suitable for all terrain.
- Portable systems can be installed very quickly.
- They are often the most cost-effective solution.
- Radio systems are very resilient—it is difficult to dig up a radio link!

The main downside of radio systems is their bandwidth limitations, typically 2 to 140 Mbps, and their use of scarce radio frequencies. Radio frequencies are a very limited resource and there is strong competition for frequency allocations in most countries. In some countries, frequencies are auctioned off to the highest bidder. In others, their allocation is tightly controlled by the regulatory authorities.

Terrestrial Radio

Terrestrial radio links require a line-of-sight path between the two locations to be linked (Figure 2.6). They are often used for long-distance traffic as an alternative to fiber. In this case the line-of-site requirement is not particularly onerous, as permanent towers can be installed to give the required height. An occasional problem can arise with permanent links whereby a new building can obstruct what was previously a clear path.

Satellite

Satellite links overcome the line-of-sight limitation of terrestrial radio and hence are often more suitable for temporary links.

Many communications satellites operate in what is known as *geostationary* orbit. They rotate around the globe once every 24 hours directly above the equator (i.e., at the same rate at which the Earth spins on its axis). In this way they appear stationary in the sky. The satellite dishes can be pointed directly at

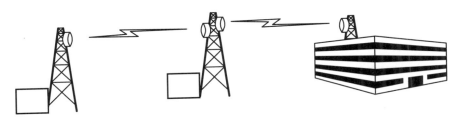

Figure 2.6 A radio link.

them without the need for elaborate tracking mechanisms. To achieve this, the satellite must be placed in orbit 36,000 km above the equator (Figure 2.7).

The satellite is used as an amplifier in the sky. It accepts information from Earth stations on the ground and relays it back to its destination.

The biggest limitation of satellite links is that they introduce a significant delay. The information has to travel up to the satellite and back; and, for geostationary satellites, this takes approximately one-quarter of a second. This delay is particularly noticeable with voice communication. Initially, satellite communication was only viable for large organizations with the necessary resources needed to build large Earth stations. With today's technology, however, miniature terminals are available, which can be cost effective for a diverse range of applications.

2.3.3 Switches, Routers, and Nodes

As the number of devices in a network grow, a stage is quickly reached whereby it is no longer practical to have a direct communication link between each of them. For example, a network of only 100 devices would require almost 5,000 links. Increasing the number of devices to 1,000 would require almost half a million links. For this reason, *switched* networks have been used since the early days of telephony.

In a typical switching scenario, individual devices have direct links to a central switch. Connections between devices are set up only when they are required. To cater to geographical dispersion, multiple switches can be used with links between them. Individual devices are connected to the nearest switch.

A *switch* or *router* is thus a device used at the junction between a number of communication links. Its function is to send information in the right direction. The term *router* is generally restricted to computer networks while *switch* is used on both voice and data networks.

The term *node* can be used to denote either device, but it is used somewhat inconsistently when comparing computer networks with voice networks. On a

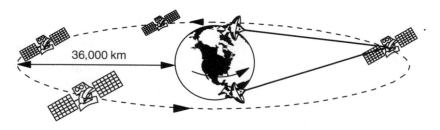

Figure 2.7 Satellites in geostationary orbit.

computer network, every computer is a node regardless of whether it performs routing or not. (Thus, on a network of PCs, every user has a node on his or her desk.) On a voice network, on the other hand, only the switches are called nodes, and the telephones (user equipment) are called *stations*.

2.3.4 Network Topology

The topology, or layout, of a network is the way in which the individual network nodes are connected together. In some cases, particularly for single-site networks, the topology is determined by the particular technology being used. With multisite networks, the optimum topology will often be a trade-off between network cost and reliability. Here the communication links will have to be rented from a service provider, and the cost will normally be dependent on both distance and bandwidth. In such cases the most cost-effective topology is often one that limits the average length and capacity of the various links. Four common topologies are shown in Figure 2.8 and are described next.

Star Topology

A star network consists of a single central node or hub, with all other nodes connected directly to it. This configuration is particularly suitable for a business with a primary location and a number of subsidiary locations. The primary location might be the business' headquarters or the data center where most of the information processing is carried out. The subsidiary locations might be branch offices or sales outlets. In such cases, star networks can be both economical and efficient, particularly where the communication requirements between one subsidiary location and another are minimal.

A star topology is also used to minimize the overall cost of WAN links, particularly for international networks. There is often intense competition between service providers to woo businesses with significant communication requirements. Some are prepared to offer significant financial incentives if a company routes all or most of their links via a specific location or *hub*. This can be a powerful economic argument for using a star topology.

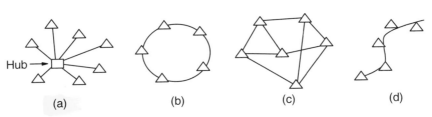

Figure 2.8 Network topologies: (a) star, (b) ring, (c) mesh, and (d) bus.

There are two potential disadvantages of star networks. The central node represents a single point of failure, because if it fails, all communication ceases. This risk can be minimized at the design stage but cannot be totally eliminated. Second, as the subsidiary nodes are not directly linked to one another, the central node must carry information for the entire network and can become a potential point of congestion. This risk can be eliminated if it is given due consideration at the design stage. For example, where the information-handling capabilities are uncertain, a flexible solution can be chosen with the capacity of the central node being readily expandable as required.

Ring Topologies

In ring networks, nodes are connected by means of direct links to form closed loops. This means that there are always two links to each node and, hence, two alternative paths for information to travel. Ring topologies are often used to provide protection against the effects of a single communication link failure. To provide this protection, many parts of the network are duplicated. For example, a main fiber and a standby fiber are often used. The topology also provides some protection against node failure. Should a single node fail, information flows between the remaining nodes need not be affected.

Ring topologies are used in some public networks, in some LANs, and in small private WANs based on *fiber distributed data interchange* (FDDI). Networks based on this topology can be difficult to modify, as two additional links are required for each new node. In addition, the nodes must be designed to carry the additional information that might have to pass through them should a failure occur. The net result is that implementing a fully protected ring network can involve significant additional expense.

Mesh Topology

Mesh networks can be considered ring networks with additional communication links installed. These networks are particularly robust, and in many cases users will be unaware if a network failure occurs. When only a limited number of additional links have been added, the topology is called a partially connected mesh. (Strictly speaking, to be classified as a mesh, there should be at least three separate links to each node.) Where each node is connected to all others, the topology is called a fully connected mesh.

Bus Topology

With this topology, nodes are linked by means of a single cable. A physical connection is made to the cable at each node, and all nodes share the available bandwidth. Bus networks require the minimum amount of cable and hence are often the cheapest solution. Furthermore, they are easy to modify, as an additional length of cable can be added at the end. There are, however, distance limitations for all bus networks that depend on several factors: protocol type, speed, and cable type. Cable segments can be combined using repeaters, but once again there are often limits as to how many repeaters can be added in series.

Bus networks are widely used for LANs and for cable television networks. Of all topologies, they have the lowest cabling costs. However, such networks are the least resilient of all. A break in the cable will generally isolate an entire section of cable and disrupt all communication. This makes them extremely vulnerable in situations where the risk of a cable break is significant (e.g., along a road).

Hybrid Topologies

Many networks have topologies that are a mixture of the foregoing categories. These are sometimes called hybrid topologies.

A Word of Caution

The topology alone does not determine the resilience of a network (its ability to survive should a failure occur). First, the communication links themselves must be considered. Physical cables will often share the same physical path, and may even be different pairs in the same cable. In such cases, all communication links connected to a node are liable to fail simultaneously. Second, the equipment at each node plays a major role. Not only must it be reliable, it must also have the capability to take appropriate action (e.g., reroute the traffic) should a failure occur.

2.4 Public Networks

A public network is a shared network that provides connectivity to many individual users. Public networks are used for a wide range of intersite communication needs and many different types are in use today. The most common example is the ordinary telephone network. Such networks may provide either switched or permanent links; they may be digital or speech-only and either high or low bandwidth. Many are optimized for certain types of information or specific applications.

2.4.1 Examples of Public Networks

Table 2.4 gives a brief definition and description of the main public network types available today. Chapters 5 through 9 of this book give more details.

Table 2.4
Public Communications Services

Name	Definition and Description
Public switched telephone network (PSTN)	The ordinary telephone network; access is normally over an analog link
Integrated services digital network (ISDN)	An enhancement of the PSTN that offers end-to-end *digital* connections and advanced services
X.25	A low-speed data network with very good global coverage but unsuitable for voice
Frame relay	A medium-speed data network ideal for interconnecting LANs and for other data applications; voice communication is possible at a slightly degraded quality
Internet	A global data network offering a variety of speeds but few (if any) guarantees of quality of service
Switched multimegabit data services (SMDS)	A high-speed data network offering speeds up to 25 Mbps; global coverage and interconnections are patchy
Asynchronous transfer mode (ATM)	A very-high-speed data network (up to 140 Mbps in 1997). The most important characteristic of ATM is its ability to mix voice, video, and data in an effective manner. Global coverage and interconnections are very patchy.
Cellular radio	Mobile network for voice and low-speed data
Leased lines	Analog or digital private circuits that can be used to build almost any kind of private network

2.4.2 Characteristics of Switched Public Networks

The switching techniques used in switched public networks can be divided into two distinct categories: those that allocate a fixed bandwidth for the duration of the call and those that allocate bandwidth only as required. Where a fixed bandwidth is allocated, the network is said to be *circuit switched*. If a network allocates bandwidth only as required, it is said to offer "bandwidth on demand" and is often referred to as "packet switched." These differences are often reflected in the charging mechanisms. Fixed-bandwidth calls tend to be charged on a duration basis, whereas those over a bandwidth on a demand network tend to be charged on the basis of the actual information transmitted.

Within these two classifications there is a wide range of options. The most significant is the maximum bandwidth available. Networks that offer bandwidth below 2 Mbps are often referred to as *narrowband.* Those offering bandwidth at speeds above this figure are often called *broadband.* Broadband switched networks are not widely available (as of 1997), and even where available their associated costs are often prohibitive. Because of this, the vast majority of switched WAN links are at speeds of 1.5 Mbps (2 Mbps in Europe) and below. Not only that, again due to cost constraints, these speeds are still considered high! In contrast, many single-site networks (LANs) now offer connectivity at speeds of 100 Mbps and above.

Another important parameter is the length of time, or delay, before information is available at the far end. There are two main sources of delay. First, it takes a finite time to establish a connection. For example, data calls over the PSTN require a modem, and the establishment of a connection between two modems can take 20 seconds or more. Second, there is the time taken for the information to transit the network. With some *store and forward* networks, information can take many minutes to arrive at their destination.

Closed User Groups

A *closed user group* (CUG) is a set of users on a switched network who can communicate with one another but not with users outside the group. They serve two functions: security (i.e., keeping intruders out) and cost control (e.g., limiting the places into which your employees can call).

CUGs are generally quite flexible:

- Users can be members of a number of CUGs.
- The restrictions can apply to incoming calls only or to outgoing calls only. Incoming restrictions meet security requirements; outgoing restrictions meet cost control requirements.

2.4.3 Access to Public Networks

The connection between the customer's premises and the public network is known as the access line. In many cases (e.g., telephony and low-speed data), the connection uses a multipair copper cable with each user having his or her own dedicated pair (or pairs) (Figure 2.9). This has significant economic advantages because the cost of laying and maintaining the cable is shared. In other cases (particularly for high-speed connections), a dedicated access line may be

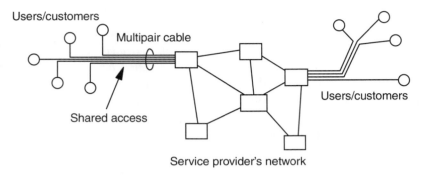

Figure 2.9 Public switched networks are shared by many users. Shared multipair cables are used for access to the network.

required. This can often be very expensive, as the costs involved will often have to be borne by just one user.

2.5 Standards

Standards play a major role in the information and communication technologies arena. They offer a number of advantages over nonstandard solutions:

- They allow products and services from diverse suppliers to be interconnected.
- They allow economies of scale, and hence lower equipment prices, by creating large markets for common products (e.g., the *Global System for Mobile Communication* [GSM]).
- Futureproofing—standardized interfaces makes it easier to add on new equipment.

Standards Bodies

A number of bodies sets standards [2] or makes recommendations that have the force of an official standard pertinent to the telecommunications industry. These bodies are listed in Table 2.5

Many other bodies play a very important role in the *writing* of standards (as opposed to official ratification). These include the ATM Forum, the Frame Relay Forum, the SMDS Interest Group, and the *Gigabit Ethernet Alliance* (GEA). These bodies have been set up by interested parties (usually vendors). In some cases they do all almost all of the work on the standard and then submit it to the appropriate standards body for ratification.

Table 2.5
Standards Bodies

Body	Typical Standard
International Telecommunications Union -Telecommunications Standardization Sector (ITU-T). This organization used to be called CCITT.	A wide range of standards for use in public communication networks (e.g., ISDN, X.25, PCM)
American National Standards Institute (ANSI)	A very broad range of standards (e.g., FDDI)
European Telecommunications Standards Institute (ETSI)	Standards for public communication networks (e.g., GSM, Euro ISDN)
Internet Engineering Task Force (IETF)	Internet standards (e.g., Internet mail, simple mail transport protocol [SMTP])
International Standards Organization (ISO)	OSI 7 layer model
Electrical Industry Association– Telecommunications Industry Association (EIA/TIA)	Electrical interfaces used for interconnecting digital equipment
Institute of Electrical and Electronics Engineers (IEEE)	LAN standards (e.g., Ethernet)

The standards-setting processes have come under much criticism, mainly associated with the length of time a standard takes to come out and excessive influence by vendors, often against the interests of users.

The alternative to using standards is to adopt a proprietary solution. This will often deliver superior performance—sometimes even at a lower cost. Proprietary solutions have the significant drawback that they lock the consumer into a particular manufacturer's products. To overcome this problem, some products are *multistandard* and can work in proprietary mode when communicating with a similar product or in standards-based mode when communicating with another vendor's product.

References

[1] ADSL Forum, "General Introduction to Copper Access Technologies," http://www.adsl.com/general_tutorial.html, 1997.

[2] Heywood, Peter, Mary Jander, Erica Roberts, and Stephen Saunders, "Standards: The Inside Story," *Data Communications International,* March 1997.

3

PBXs

3.1 Introduction

A *private branch exchange* (PBX) is a private telephone switch located on a business premises or factory. It has connections to both internal *lines* and external *trunks*. The internal lines are used for extension telephones on the premises, while the external trunks link it to the public telephone network. The extension telephones can call one another without using the public network. They can also make and receive calls to and from the public network using the external trunks. PBXs in different buildings can be linked together with special trunks, sometimes referred to as tie lines, to form a private network. PBX connections are illustrated in Figure 3.1.

Using a PBX, an organization can meet all of its internal telephone needs without having to pay call charges. It can minimize the number of telephone lines that it rents from the public telephone operator. It can better manage the

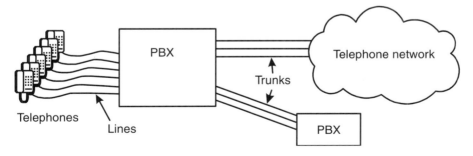

Figure 3.1 PBX connections.

calls that must use the public network through the use of barring features and detailed call logging. It can advertise a single telephone number for the business and give advanced telephone features to the employees.

In recent years many organizations have established *call centers* to stream-line their dealings with customers. A call center is a place where people are em-ployed to interface with customers over the phone. The majority of call centers deal with sales, technical support, or information services. At the heart of every call center is a feature-rich PBX system capable of handling heavy call traffic and preferably linked with the center's computer system.

3.1.1 A Key Component in any Communications Strategy

In contrast to email, fax, or letter, voice communication is immediate, personal, and the most widely available. In general, email and fax are not widely available to the public (although things are rapidly changing in this area) and letter writ-ing is part of the culture of yesterday. Voice is therefore the most important form of communication in almost every industry today.

Great emphasis has always been placed on the reliability and availability of a telephone service, be it public or private. Today's PBXs are no exception.

3.1.2 PBXs, Key Systems, and Hybrids

In the past, there was a strong distinction between PBXs and smaller switches called *key systems.* The original key system was a small switch that could concen-trate a number of extension phones onto a smaller number of external trunks. Individual trunk lines were accessed with separate keys.

These systems had very limited functionality when compared to a PBX. Key systems have all but disappeared; they have been replaced by micro-processor-controlled switches with a functionality approaching that of a PBX. These switches are often called *hybrids* because they have some of the charac-teristics of both PBXs and key systems. For the rest of this chapter, we will use the term *PBX* to refer to all types of systems.

3.1.3 PBXs Versus the PSTN

Later in this book we will see how an organization could use the PSTN to pro-vide for all of its telephone needs (through the use of Centrex and *virtual private network* [VPN] services). By and large, however, it makes a lot of sense for an organization to use PBXs in all but its smallest sites. The main benefits of such an approach are as follows:

- More features, such as call waiting, call forwarding, and call pickup;
- Better utilization of public exchange lines;
- Fewer public exchange lines to rent;
- "Free" internal calls.

3.2 PBX Structure

A PBX consists of a switch connected to a number of *ports* which in turn are connected to *stations* (telephones) or trunks (Figure 3.2). Ports are referred to as *line interfaces* or *trunk interfaces* depending on whether they are connected to a phone or a trunk. Physically, one or more ports are built onto a printed circuit board that is slotted into the PBX chassis. The ports and switching module are monitored and controlled by a processor (computer).

3.2.1 Analog Versus Digital

Modern PBXs are electronic and can be either *analog* or *digital* (see Section 2.2). Digital PBXs convert the voice into digital signals either within the telephone instrument or at the point where the telephone line connects to the PBX (the line interface). They do not usually improve the voice quality on internal calls, but digital PBXs offer the following advantages:

- They can be linked together into an all-digital private network. Speech quality on long-distance connections is maintained if all of the intermediate links are digital. There will also be significant economies of

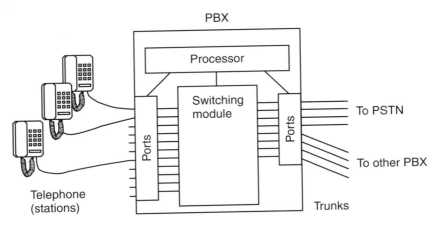

Figure 3.2 Structure of a PBX.

scale if the PBXs are large and require high-capacity digital trunks (1.5 or 2 Mbps) between them.

- Digital PBXs often support ISDN connections to a public network, thus giving the possibility of advanced services to each station. ISDN is dealt with in greater detail in Chapter 6.

- Most digital PBXs can support data communication at 64 Kbps.

3.2.2 Telephones (Stations)

The telephones used on a PBX are often referred to as *stations*. The most flexible of today's PBXs allow a variety of telephone types to be connected (i.e., both *proprietary phones* and *standard telephones*).

Proprietary phones are manufactured by a PBX supplier and only inter-work with that supplier's PBX. They are often called *feature phones* and are more expensive but much easier to use than standard phones. They allow features to be accessed by a single key press, rather than a string of digits preceded by the * or # keys. This ease of use means that important features, such as call transfer, actually get used, and a good impression is made on the caller.

Other features such as hands-free operation, visual display of who is calling, and indicator lights to show calls on hold or waiting for an answer can also be included in feature phones. Given that the telephone is such a major part of some people's jobs, it often makes sense to invest the extra money on a feature phone to improve the quality and efficiency of that part of their work.

Standard phones are manufactured for use on public networks or on generic PBXs. They can be purchased from a large range of suppliers and are consequently much cheaper. Standard *digital* phones are phones manufactured to an ISDN standard. At the time of writing there are still a number of variations on the ISDN standard in different parts of the world, but this situation is expected to stabilize with consequent reduction in the cost of this type of phone.

Care should be taken, when seeking quotations for a PBX, to find out the cost of proprietary feature phones, even though you may not want to purchase any in the short term. A sales tactic with some suppliers is to quote a very competitive price for a system with standard phones, but to subsequently charge a very high price for feature phones (which cannot be sourced elsewhere once you have fixed on a model of PBX).

Digital phones can only be connected to digital PBXs, whereas an analog phone can be connected to either (provided the digital PBX has analog ports).

A most important station on the PBX is the operator's console. This station will have a lot of functions unavailable to the standard user.

3.2.3 Cabling

Twisted-pair cabling is used between the PBX and the telephones. The cable is wired in a star configuration, each station having a dedicated connection back to the PBX. This connection will, at a minimum, be a single pair of wires. Some PBXs, however, require two or more pairs per station. This is an important parameter when replacing a PBX. If the new PBX requires more pairs per station, then you may not have enough cable in place to support it.

The type of cable used is referred to as voice-grade cable. In large installations, the PBX ports will be connected to a wiring frame for flexibility. From there the wiring is distributed using large multipair cables, fanning out into smaller cables as you get closer to the phones.

The *structured cabling* system for LANs (see Section 4.4.2) is often used for the both the LAN and the phone system. In this case, the phone system may use data-grade cable. This is more costly but is more flexible because phone and LAN connections will be interchangeable.

3.2.4 Switching

The primary function of a PBX is to switch calls between two parties. This means making a connection, inside the PBX, between two ports (Figure 3.3). This function is carried out by a switching module or switching matrix.

The switching module determines whether the PBX is analog or digital (i.e., an analog switching module implies an analog PBX, and a digital switching module implies a digital PBX).

Some PBX switches are designed on the basis that not everybody will want to use the phone at the same time. For economic reasons, they allow a small probability that calls will be blocked within the switching matrix. This means

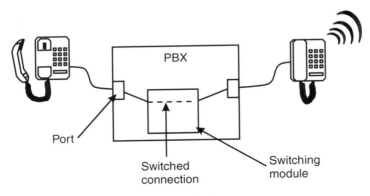

Figure 3.3 A switched connection between two telephones.

that during peak periods you may get busy tone when you call a station that is actually free. Typically, a 1 in 500 probability of call blocking during the busiest hour of the day is considered acceptable. Many modern switching modules are *nonblocking*. This means that the maximum number of possible simultaneous calls can in fact be made at any time. The relatively low cost of modern digital switching devices means that nonblocking switches are now much more common.

3.2.5 Processor

The processor in a PBX is responsible for routing calls to their destination. To do this, it must be loaded with software programs (written by the supplier) and data, in the form of tables, listing where everything is connected, how calls should be routed, and what facilities and features each station is allowed to use.

In small PBXs, the software and most of the data is stored in *read-only memory* (ROM). Some of the data, such as trunk barring and abbreviated dialing, must be stored in *random-access memory* (RAM), or read-write memory, so that the owner can change them as he or she sees fit. This data must be protected by a small battery normally located inside the PBX.

Large PBXs generally allow more flexibility. All of the software and data is loaded up from disk drives into RAM on power-up, rather like a PC. This allows software updates to be distributed on diskettes rather than on ROM chips.

The suppliers of large PBX systems are constantly updating their software with new facilities or fixes for problems in older versions. Although this means that things are always improving, it does have an associated cost. Furthermore, most suppliers are unwilling to support software releases which they consider to be too old. Customers are thus left with the option of going it alone or paying for the software updates.

Most large PBX systems come with a standard set of features plus optional extras, which can be added by means of a software change plus perhaps a hardware addition or modification.

3.2.6 Trunks

The term *trunk* is used to refer to a connection to another exchange, either private or public. Links to other third-party equipment are also called trunks. Examples include *music trunks* for music-on-hold, *paging trunks,* and *recorded announcement trunks.*

Exchange Trunks

The trunks that connect the PBX to the nearest telephone exchange (central office in the U.S.) can be incoming-only, outgoing-only, or both incoming and outgoing. They are very similar to the ordinary telephone lines that any household would have, but they normally have additional features appropriate to a PBX.

Trunks can be either analog or digital. Digital trunks can be ISDN or non-ISDN. Non-ISDN digital trunks are referred to as *switched digital* lines; they were introduced as a stop-gap arrangement prior to ISDN, and they are often more expensive than ISDN lines.

Large digital PBXs can be connected to a public exchange using E1 (or T1 in the U.S.) leased lines, giving the equivalent of 30 (or 24) digital lines in one connection. Again there is the option of ISDN or non-ISDN. The facilities available on exchange trunks will be discussed in more detail in Section 3.4.3.

Inter-PBX Trunks

These are leased lines that connect two or more PBXs together to form a private network. They are often called *tie lines.* Table 3.1 lists the possible types of inter-PBX trunks.

Speech compression (or low bit-rate voice coding) can be used to increase the capacity of a digital line (by a factor of seven or more) as described in Sections 11.3.1 and 11.3.4.

Each type of trunk requires a different type of trunk interface. The trunk interface is a circuit board that slots into the PBX. In general, the E1 or T1 interfaces are the best value per circuit, but you need sufficient traffic to justify the large capacity. Analog interfaces for digital PBXs are the most expensive.

Table 3.1
Types of Inter-PBX Trunks

Type	Number of Voice Channels
Analog 2 wire	1
Analog 4 wire	1
Digital 64 Kbps (56 Kbps in parts of the U.S.)	1
Digital E1 or 2 Mbps in Europe	30
Digital T1 or 1.54 Mbps in the U.S.	24

In addition to leasing a line from a public carrier, there is also the option of using a satellite link and leasing capacity from a satellite consortium.

3.3 PBX Networking

PBX networking means connecting two or more PBXs together using leased lines to form a private network. A private PBX network gives the owners greater control over their communication costs, and in most cases helps to reduce these costs. This is normally the primary reason for installing a PBX network.

PBX networking has been possible since the 1960s; however, it wasn't until the introduction of software-controlled PBXs and sophisticated inter-PBX signaling systems that the full power of networking was enabled. PBX networking today can offer:

- Feature transparency across the network (this means that features such as call waiting, call transfer, and callback can be used between extensions on different PBXs);

- The ability to share resources such as operators, voice-mail systems, and trunks;

- The ability to duplicate resources to minimize the possibility of a total system breakdown.

PBX networks are normally built to link a number of remote premises together; however, it is also possible to build a PBX network on a large premises or campus by installing separate PBXs around the premises. This option will generally be more expensive than that of installing one single large PBX, but it will increase the reliability of the communications as a whole. For example, it is possible to spread PSTN trunks over more than one PBX, and critical areas can be served with stations from more than one PBX.

3.3.1 Break In and Break Out

The economic advantage of a PBX network can be further enhanced if it is possible to "break in" and "break out" of the network. This means making long-distance calls, to or from "off-net" phones, using the private network to make the long-distance part of the connection. Two possible connections are illustrated in Figures 3.4 and 3.5.

Break out is very common; break in, however, requires the facility for an off-net phone to get a dial tone from the PBX. This facility is called *direct in-*

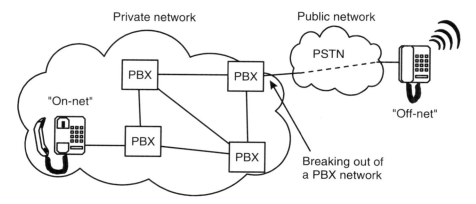

Figure 3.4 Breaking out of a PBX network.

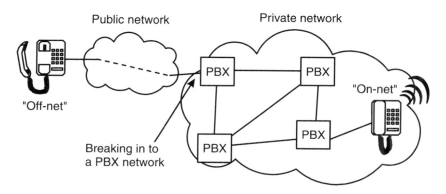

Figure 3.5 Breaking in to a PBX network.

ward system access (DISA). DISA is further discussed in Section 3.5.5. The use of DISA on a PBX entails a risk of toll fraud, whereby unauthorized callers make long-distance calls via your PBX (see Section 3.7).

3.3.2 Mixed Networks

The PBXs in a large network are often different models and may even come from different manufacturers. The differences are due to dissimilarities in the size of sites, different dates of purchase of the PBXs, and tendering processes. Such a network will require special arrangements to allow the different PBXs to interwork. For example *analog to digital converters* (A-to-D converters) will be required between analog and digital PBXs.

These converters can be integrated into the trunk interface cards on the digital PBX if the trunk (leased line) is analog (Figure 3.6).

If the trunk is digital, the A-to-D converters should be located next to the analog PBX (Figure 3.7). This arrangement is better if you wish to develop your network into an all-digital network at some point in the future.

3.3.3 Signaling Between PBXs

PBXs must be able to "talk" to one another to establish and terminate calls. First, they must be able to send dialed digits to a distant PBX so that it can switch the call to its correct destination. They must also be able to transfer such information as the answer and hang-up signal.

Tone-Based Systems

A variety of signaling systems, most of them involving the use of audio tones and pulses, have been developed for public telephone networks and have been adapted for use in PBX networks. These signaling systems use the speech circuits of a particular call to transmit information about that call.

These audio-signaling systems are quite limited. First, few different signals are available so it is difficult to add new features. Second, the majority of the signals can only be sent during the call-establishment phase and not during the speech phase, making it difficult to add features that can be used during an established call.

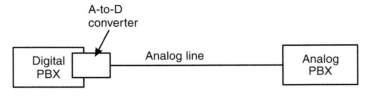

Figure 3.6 Analog trunk linking analog and digital PBXs.

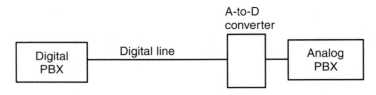

Figure 3.7 Digital trunk linking analog and digital PBXs.

Message-Based Systems

The advent of processor-controlled exchanges allows the development of signaling systems based on computer messages exchanged over dedicated signaling links. Message-based signaling systems can have an enormous number of different signal meanings, and signals can be exchanged at any stage during the call.

Message-based signaling systems are often called *common channel signaling* systems because they use a common dedicated signaling link for all calls rather than using the individual speech circuits. They are at the heart of many of today's advanced PBX networking features. "QSIG" and "DPNSS" are examples of message-based signaling systems used between PBXs.

3.4 General PBX Features

3.4.1 User Features

Even the smallest modern PBX comes with a long list of features, many of which will only be of interest to a small fraction of businesses. Furthermore, many useful features go unused because of lack of user awareness and difficulties associated with remembering how to use the features. Thus, when evaluating a PBX's features, it is not a simple question of looking for the most. Rather it is a question of finding the PBX that meets your needs in the most user-friendly and cost-effective way.

Useful user features include: abbreviated dialing, automatic call back, call transfer, call divert or call forward, conference calls, call waiting (an indication, either visual or aural, that another person is calling you), and call pickup, where you can answer a call that is ringing on a nearby station.

3.4.2 System Features

System features are somewhat hidden to users. These include:

- *Call hunt:* Call hunting is used where calls to a particular department can be answered equally well by anybody in that department. All the extensions in that department can be made part of a *hunt chain.* If an extension is busy, an incoming call will be automatically diverted to the next free extension in the chain. Call hunting is a poor alternative to *automatic call distribution* (ACD) discussed in Section 3.5.1.

- *Distinctive ringing:* This allows, for example, external calls (calls from the PSTN) to ring differently from internal calls.

- *Toll restriction:* Certain stations can be barred from making long-distance or international calls.

- *Night service:* Normally all incoming calls from the PSTN are routed to the attendant console. Night service allows these calls to be routed to some other extension or simply to a large bell. This can be done at night or at any time when the attendant is not on duty.

- *Power fail transfer:* When the power fails, a standby battery normally takes over. If the failure is prolonged and the battery expires (or if there is no battery), the system will connect the PSTN trunks to specific extensions. These extensions must not be feature phones, which cannot interwork directly with the PSTN.

- *Direct inward dialing* (DID):[1] This allows outsiders to dial an extension directly without the intervention of the attendant (see also Section 3.4.3).

- *Alternate route selection:* If all the trunks on a particular route are busy the PBX can try another route that eventually leads to the same destination because there is a mesh network (see Figure 3.8).

- *Least-cost routing* (LCR): This allows the PBX administrator to program the PBX to automatically choose the most economic route available, based on the number dialed and the time of day. This feature is normally combined with alternative routing to ensure that if all the trunks on the least-cost route are busy, the call will be diverted to the next most economic route.

- *Detailed billing:* Most PBXs have the ability to output call details to a printer or to a general-purpose computer, such as a PC. Software that can sort this data and calculate charges for individual calls or collate

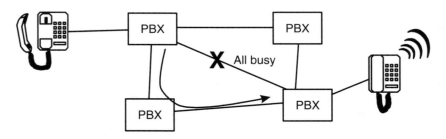

Figure 3.8 Alternative routing.

---

1. The term *direct dial in* (DDI) is also used.

charges for a particular extension or department over a period of time may be available from the PBX manufacturer or from a third party.

Billing systems such as these serve the following purposes:

- They reduce employee abuse of the phone system. Even the mere announcement that such a system has been installed can reduce the phone bill by about 25%. (A printer sitting on the receptionist's desk or a "please explain" note from the accountant will have an even greater impact.)
- They are very popular in hotels for passing on call charges to customers.
- They can be used for allocating costs to different business units.

3.4.3 PSTN Facilities

Telephone trunks that attach a PBX to the PSTN can be exactly the same as the ordinary PSTN lines used in residential installations. In most cases, however, the PBX owner will request additional features on the trunks. Each of theses facilities must be requested from the telephone company that supplies the trunks.

Hunting

This is a must for PBXs with more than three or four trunks. It allows the PBX owner to advertise a single telephone number for his business. When a person dials the main number for a PBX, the PSTN exchange connected to that PBX will hunt through all the trunks connected to the hunt group to find a free one (see Figure 3.9).

PBXs with only two trunks may be more effective without hunting. That way, one line is kept free for outgoing calls. Similarly, with larger PBXs,

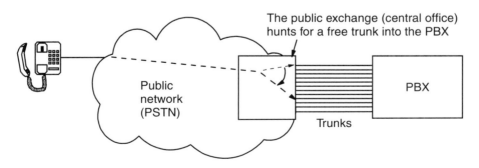

The public exchange (central office) hunts for a free trunk into the PBX

Public network (PSTN)

PBX

Trunks

Figure 3.9 Hunting.

some trunks can be excluded from the hunt group. These trunks are effectively outgoing-only.

Meter Pulses

These are sent from the PSTN to the PBX every time the PSTN steps its charge meter. This improves the accuracy of the PBX's call logging. They are sent on trunks that are engaged on outgoing calls. They are sent by means of 12- or 16-kHz pulses, which are inaudible to the average human ear. On ISDN trunks, they can be sent as a message on the signaling channel.

Direct Inward Dialing

DID allows callers to the PBX to dial a short prefix followed by an extension number (e.g., 701–7125, where 701 is the prefix and 7125 is the extension number) and get through to that extension without the intervention of an attendant. It substantially reduces the workload of the attendant.

3.5 Advanced Features

3.5.1 Automatic Call Distribution

This is an advanced form of hunting that allows you to distribute incoming calls more evenly among attendants. Calls are forwarded to the attendant who has been free the longest or to a free attendant who has dealt with the least number of calls. Attendants log in and out of the ACD system, thus avoiding calls ringing at unattended stations. Finally, an ACD system can supply statistics on call handling.

ACD is used in call centers and other situations where a large number of people are answering calls relating to the same topic. Examples include telesales, help desks, service centers, and inquiry services.

3.5.2 Wireless Capability

A wireless PBX allows the users carry a mobile extension phone with them around the premises with the capability of both making and receiving calls. Mobile stations can exist alongside traditional fixed extensions, and in most cases wireless capability can be added on to an existing PBX with most users continuing to use their fixed extensions.

Wireless PBX coverage is based on micro cells. Base stations consisting of a small wall-mounted antenna with a limited coverage area, of, say, 50m radius, are strategically placed around the premises. The coverage area of each base station is called a cell. Most systems allow calls to continue uninterrupted as you move from one cell to another.

The radio frequencies used do not generally require a special license, provided the equipment meets the required standards.

3.5.3 Voice Messaging

Voice messaging systems allow a PBX to automatically play prerecorded messages in response to the actions of the caller. For example, I call the airport to inquire about the arrival time of a particular flight. My call is answered by the messaging system, and I am presented with a *voice menu*—a prerecorded greeting inviting me to press "1" for British Airways, "2" for Virgin . . . and "6" for none of the above. After pressing "1," I am further invited to type the numeric part of the flight number and on doing so I am told the arrival time. All of the messages are prerecorded, so the system can run without human intervention for most inquiries. The systems can be programmed to put the caller through to a human attendant by dialing a particular digit. This type of system is often called *interactive voice response* (IVR).

These systems work best if the caller is using a touchtone phone (i.e., a phone that dials out using different tones for each digit). The alternative dialing arrangement is called pulse dialing, where, for example, three pulses represent the digit "3" and so on. Pulse phones do not work well with voice-messaging systems because the pulses are suppressed in the telephone network. They arrive at the far end as a fairly distorted series of clicks. Equipment that can convert these clicks back into tones is available, but it is seldom more than 90% reliable. In particular, the digit "1" is very hard to detect correctly. The best way to handle calls from pulse phones is to divert them to a live attendant. This diversion can be triggered if the system does not detect any tones from the caller within, say, 10 seconds. Another alternative, which will become more popular as the technology matures, is to use speech-recognition equipment allowing the users to speak their requests.

3.5.4 Voice Mail

Installing a voice-mail system is the equivalent of giving everyone on the PBX a personal answering machine. In this case, they get a voice mailbox instead. They can retrieve their messages from any phone, even one not connected to the PBX, by dialing in their extension number followed by a password. They can also record their own personalized greetings. In addition to all of the facilities that you get with an answering machine, the following features of voice mail make it more attractive:

- Messages are digitized and stored on magnetic disk. Thus, the sound quality is often superior to answering machines, and the access time to messages appears instantaneous because you do not have to rewind or fast forward a magnetic tape.
- The voice-mail system is centralized, and the cost per station is often less than for individual answer machines.
- Messages you receive can be forwarded to other voice mailboxes.
- Voice-mail systems allow you to compose and send messages to a mailing list (of all your staff, for example).

3.5.5 Direct Inward System Access

DISA is an alternative method to DID that allows outside callers to dial directly to the stations on the PBX. In this case, the caller dials the DISA number (which may be different from the main number advertised for the PBX) and receives back a dial tone or a recorded message from the PBX. The caller can now dial the desired extension number.

This system has the advantage over DID that it doesn't require any intervention from the telephone company, and you are not required to rent blocks of numbers. DISA does, however, require that anyone dialing in has a touchtone phone. If the caller doesn't have a touchtone phone, a good DISA systems will transfer the call to an attendant after a fixed period of time.

DISA systems can be combined with voice menus so that instead of receiving a dial tone, the caller gets some assistance for what to do next by means of voice prompts.

3.5.6 Data Communications Over PBX

Many PBXs offer the possibility of using the PBX as a switch for data calls. Typical speeds range from 9,600 bps to 64 Kbps, the latter being common on digital PBXs. These speeds are fine for connecting terminals to a host computer but are no match for LAN speeds when it comes to connecting PCs to file servers. In this case, much greater bandwidth is required, especially when loading application software from a file server.

In many cases the data communication capabilities of PBXs go unused because the owners use a LAN for their data communication. Some interesting applications for data over a PBX include:

- The sharing of a modem pool (which can also be achieved using a LAN);
- The simultaneous switching of voice and data—imagine you are dealing with a customer query; you have all of their details up on a screen in front of you, and you need to transfer the call to another person. It would help matters if you could transfer the data connection along with the voice call so that the recipient will be instantly aware of the identity and details of the caller. This can easily be achieved if you are using the PBX to route your data connections as well as the voice connections. *Computer telephony integration* (CTI) will allow the same result to be achieved even if the data is routed over a LAN.

3.5.7 Computer Telephony Integration

Many modern PBXs can be connected to general-purpose computers using a link cable known as a CTI link. This allows the PBX processor to communicate with a third-party computer system to provide a new set of integrated applications.

Examples include:

- Databases that react to the calling line identification of an incoming call and "pop up" information relating to the caller (so-called *screen-pops*);
- Databases that react to the DID digits coming from the PSTN;
- Call management systems—make or transfer calls by selecting a person's name on a computerized phone book;
- Voice-mail management systems—view a list of voice messages including their origin and duration on the PC screen and select which one to listen to first. The messages are listened to using the telephone instrument or the PC speaker.

Some implementations of CTI allow third-party software developers to develop applications such as ACD and least-cost routing, which are not PBX specific.

Software companies such as Novell and Microsoft have written *application program interfaces* (APIs). The idea is to present the third-party application developer with a standard set of commands and event indicators that will enable the developer to write a CTI application that is completely independent of the type of PBX, provided the PBX manufacturer supports that particular API.

Novell calls its product *Telephony Services Application Program Interface* (TSAPI). In this case the PBX is linked to a server and a client-server application logically links the user's PC with his or her phone. The third-party developers must develop software modules for both the PC (client) and the server (Figure 3.10).

Microsoft has a product called *Telephony Application Program Interface* (TAPI). In this case the telephone instrument is physically linked to the PC (Figure 3.11). Third-party software is developed for the PC alone. In this case the CTI application has less control over the PBX. It can set up and transfer calls but it cannot make routing decisions, nor can it implement ACD functions. It is best suited to small installations.

3.5.8 ISDN Capability

Digital PBXs that can connect to ISDN lines can make use of the many advanced features of ISDN, such as video telephony, high-speed data communication, and calling line identity (caller ID).

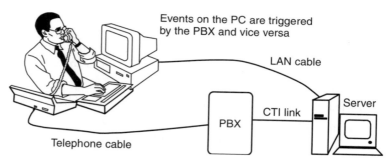

Figure 3.10 CTI link linking PBX and server.

Figure 3.11 CTI link linking PC to telephone.

3.6 Network Management

Network management means the management of all of the resources in a network. In the case of a PBX network, this means the individual PBXs (often referred to as the nodes), the inter-PBX trunks, the PSTN trunks, and the extensions. The normal functions of management include resource allocation, fault monitoring, fault localization, and repair.

It makes a lot of sense to be able to carry out as much of this work from a single location as possible, so modern PBXs have the possibility of passing alarm information over the signaling network to a maintenance center. It should also be possible to make administrative changes and software updates remotely. This can be done over the signaling network or via a dialup connection to a PBXs processor. Dialup connections are quite common because PBX maintenance is often contracted to a third party. It is essential that precautions are taken to prevent fraudsters dialing into your management system and setting themselves up as valid DISA users with international dialing rights.

3.6.1 Feature Administration

The features on even the smallest PBX need to be administered. For example, somebody must program the system, indicating which extensions are barred from international dialing. If somebody moves from one office to another, the system administrator must assign his or her existing extension number to the new phone. This work should be allocated to a trustworthy member of staff or to a trusted external contractor.

On small PBXs, features are often administered from a designated phone by keying in sequences of digits. On larger systems, administration is carried out using a keyboard and screen. Some of these systems require the administrator to enter somewhat cryptic commands. Thankfully, however, more and more PBX manufacturers and third parties are offering graphical tools for system administration.

3.7 PBX Security

PBXs are vulnerable to unauthorized use both internally and externally (i.e., unauthorized use of the PBX trunks to make long-distance or international calls—so called *toll fraud*). Examples of where this situation can occur are as follows:

- Employees abusing their rights.

- Employees, who are restricted from long-distance dialing, using stations belonging to employees with no restrictions. Phones attached to fax machines are particularly popular in this instance.

- A company allows employees who are away from the office to make long-distance calls via the company's PBX. To do this, the employee calls into the PBX and, by dialing a password, receives a second dial tone from the DISA system in the PBX. This second dial tone allows the employee to make long-distance calls over the PSTN. These calls will then be billed to the company. A fraudster could gain access to these passwords by looking over the shoulder of the employee as he or she makes a call from a public phone or by tapping the phone line and recording the tones on the line.

- A security hole can exist in voice-mail systems if users are allowed to program an alternative *divert number* into the system. (A divert number is used to allow a call to be diverted to, say, the recipient's mobile phone by having the caller press "0.") In this case the fraudster finds an extension number with an insecure password (a frightening number of PBX users have their password equal to their extension number). The fraudster then logs in to that user's voice mail, changes the call divert number to the number the fraudster is interested in, and diverts the next call to the desired number.

Fraud can be protected against by:

- Restricting trunk access to working hours;
- Assigning access codes to long-distance calls (i.e., the caller has to dial a secret prefix before all long-distance calls);
- Assigning access codes to each station;
- Ensuring that all passwords/access codes are changed regularly and are not easily guessed (e.g., access codes should never be the same as the extension number);
- Disabling system features, such as DISA and diversion to outside lines, that are known to pose security risks. The PBX vendor should be able to advise on this.

A balance must be struck between security and the needs of the business. If a feature has a high security risk and a high business value, it can be given to selected employees on the basis that they undertake to use secure passwords. Additionally, the PBX supplier may be able to provide tools for monitoring usage of these facilities and reporting abnormal events. It may even be necessary to automatically close down an account following abnormal activity. Much telephone fraud occurs after hours and on weekends when the system administrator may not be around to take action.

Alternative solutions such as providing certain employees with mobile phones rather than allowing them trunk access through the PBX should be considered. The mobile phone charges may be a small price to pay for protection against toll fraud. Remember that the professional fraudsters engage in selling international services and can run up very large bills over a short period of time. For example, a professional fraudster could make 8 hours × 60 minutes × 2 days × $0.5 × 12 trunks = $5,760 worth of international calls over a single weekend via a 24-trunk PBX. (Note that two trunks are required for every call via the PBX.)

3.8 PBX Dimensioning

The term *dimensioning* refers to calculating the correct amount of equipment required in a particular situation. The terms *sizing, traffic engineering,* and *engineering* refer to the same thing. Dimensioning a PBX means answering questions like:

- How many extensions are required—now, during its lifetime
- How many exchange lines are required—now, during its lifetime
- How many inter-PBX trunks are required—now and in the future
- What capacity of processor is required—now and in the future
- What size power supply is required—now and in the future

3.8.1 Number of Extensions

The first question concerns the number of telephone instruments to be connected to the system. In many cases this will be a straightforward count of the number of employees at a particular location.

3.8.2 Number of Trunks

The next two questions concern the number of trunks required. (You may recall that trunks are used to connect the PBX with the outside world.) These figures are more difficult to estimate because they depend not only on the number of employees but also on how much they use the telephone and how this usage breaks down between different types of call (internal calls, interbranch calls, external calls).

The best way to obtain this information is to measure it. This is possible if, for example, you are replacing an existing PBX system and you are able to take measurements from the old system. If that is not possible you may be able to make comparisons with another branch of the organization or with another similar organization. Your PBX supplier may be able to help you here.

3.8.3 Traffic (Erlangs and CCS)

Telephone usage is called traffic. Traffic is measured in *Erlangs* or in *centi call seconds* (CCS). One Erlang of traffic means that on average during the measurement period (typically one hour) one call is in progress. During that hour there may have been times when no call was in progress and other times when two, three, or four calls were in progress simultaneously, but on average there was only one call.

A CCS is a different unit of measurement for the same thing. One CCS implies 100 *call seconds* during an hour. There are 36 CCSs in an Erlang.

PBXs must be dimensioned for operation during peak calling periods—the busy hours. Traffic should be measured at these times and, ideally, averaged over a period of weeks. Many large PBXs can be programmed to measure traffic over a number of weeks. The output is called a *traffic count* and may be on paper or on disk. It is also possible to convert call logging information into traffic information if a separate mechanism for making traffic counts does not exist. Finally it may be possible to get the telephone company to supply you with traffic information for your PSTN trunks.

The results of the traffic counts can be used to estimate the number of trunks required on each route. *Traffic tables* are used to determine the number of trunks required to carry that traffic at a given *grade of service* (GOS). A typical GOS for a PBX would be one blocked call attempt in 50 (1:50).

As you can see from the traffic table (Table 3.2), economies of scale can be gained by concentrating a large volume of traffic on any route. Routes with less than four or five trunks are very inefficient in terms of trunk utilization. This problem can be overcome by using *overflow routing*. A small route is dimensioned for a very poor grade of service (1:10) but a good circuit utilization. The failed attempts are given the chance to try an alternative route before the caller is given busy tone. Provided the second route is generously dimensioned, the callers do not suffer from the low GOS of the first route.

Table 3.2
Traffic Table

Number of Circuits	Traffic in Erlangs Carried at GOS = 1:10	Traffic in Erlangs Carried at GOS = 1:50	Traffic in Erlangs Carried at GOS = 1:100
1	0.11	0.02	0.00
2	0.53	0.21	0.15
3	1.10	0.57	0.44
5	2.44	1.55	1.28
10	6.22	4.61	4.14
24	17.79	15.06	14.08
30	23.22	19.86	18.75
48	39.36	34.86	33.36
60	50.31	45.19	43.47

Source: [1]. © 1994 Roger L. Freeman. Reprinted by permission of John Wiley & Sons, Inc.

3.8.4 Processor Capacity

Processor capacity is determined by the number of call attempts as opposed to the traffic. One of the specification figures quoted for PBX processors is the number of *busy hour call attempts* (BHCA) they can handle. If however you implement sophisticated call handling features such as ACD and LCR, you will load your processor much more heavily for the same volume of call attempts.

3.8.5 Power Supply Requirements

Most PBXs require a 50V direct current (dc) supply. For large PBXs, this will often be supplied from a rectifier which converts alternating current (ac)[2] to dc. Batteries may be used as standby in case of a mains failure. Lead-acid batteries are often used. These are extremely heavy, and special consideration of floor strength is required in the case of large installations. Batteries can also pose a number of safety hazards:

- They can emit toxic, explosive gas (e.g., hydrogen) and thus require adequate ventilation.

---

2. As supplied from the electricity company.

- They contain acid, which can spill out if the battery is damaged or knocked over.
- They are capable of delivering very large current, which can cause a minor explosion and possibly start a fire if metal tools fall across the battery terminals. This adds to the dangers posed by the release of hydrogen gas.

The danger from gas and acid can be greatly reduced if recombination (sealed) batteries are used. Sealed batteries have no holes for acid to spill out, and during normal operation the gasses are kept within the battery where they recombine maintaining the acidic level. A valve allows excess gas to escape during heavy discharging and charging.

Reference

[1]   Freeman, Roger L., *Reference Manual for Telecommunications Engineering*, 2nd Edition, New York: John Wiley & Sons, 1993.

4

Local Area Networks

4.1 Introduction

The use of LANs to interconnect computer resources has become a fact of life in most businesses.

A LAN is a communication system that interconnects computer equipment in a single building or site. When several sites across the country are interconnected it is referred to as a WAN.

A typical LAN consists of a cabling system, workstations, servers, printers, and software. The cabling system will generally consist of coaxial, twisted-pair, optic-fiber, or some combination of these cable types. The workstations are the user's computers and will usually be PCs, though dumb terminals and UNIX workstations are also common. The servers can be PCs, minicomputers, or mainframes. The software can be classified as networking software and application software.

4.2 Benefits of a LAN

4.2.1 Connectivity

LANs can be used to connect all sorts of computers together, from the largest mainframes down to the simplest dumb terminals. The vast majority of LANs today are used to interconnect PCs. The most common reason for installing a LAN is to enable a number of people to work on the same information concurrently. For example, general managers can get up-to-date sales figures on their computer screens by accessing the same database their staff use to raise the invoices, and more than one person can process invoices at the same time.

4.2.2 Resource Sharing

LANs allow the sharing of expensive resources. Laser printers, color printers, processors, disk drives, CD-ROM drives, fax servers, Internet connections, and modems can all be shared on a LAN (Figure 4.1). Software licenses can also be shared. For example, on a LAN with 20 PCs, it may only be necessary to purchase 10 licenses for a particular piece of software, such as a spreadsheet program. However, this must be done on the basis that not more than 10 people will need to use the software at any one time.

4.2.3 Information Sharing—The Key Benefit

It is the sharing of information rather than the sharing of any other resource that has made LANs so popular. Some examples will make this more apparent:

- You can work on a particular document from any PC connected to the LAN, regardless of which PC was used to create the document.
- Concurrent access to databases means that information is up to date. This is essential in real-time applications, such as reservation systems.
- Concurrent access to files means that more than one person can consult a particular file at the same time. This is not possible in traditional paper-based filing systems.
- Messages can be sent to colleagues, broadcast to all users, or broadcast to a particular group of users.

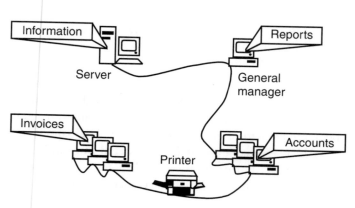

Figure 4.1 A LAN allows sharing of information and resources.

- If an email system is in use, files can be attached to mail messages. This allows users to share documents, spreadsheets, or graphics on a needs basis without having to use floppy disks.

- Workflow applications such as *Lotus Notes* make use of the above information-sharing possibilities to allow business processes to be more streamlined and structured. The structuring occurs because the process is defined in software and will be the same for everyone, the stream-lining occurs because everyone has access to pertinent information online.

4.2.4 File and Record Locking

A problem can occur if more than one person has concurrent access to a document, graphic, or database. Consider what will happen if two people decide to edit the same information concurrently. One person's editing could overwrite the other's. This situation is overcome by some sort of *locking* mechanism.

The simplest form of locking is called *file locking*. This means that the first person who opens a file is allowed to make changes and save them. Anyone else who tries to open the same file will receive a message saying that the file is *in use* but that they may open it in read-only mode. If this second user opens the file in this mode he or she will be prevented from making changes unless he or she makes a copy of the file with a different name or in a different storage area.

In a database system, a more sophisticated form of locking, called *record locking*, is possible. This allows many users work on the same database with the restriction that only one user may edit a particular record at a time.

4.3 LAN Types

4.3.1 Ethernet and Token Ring

There are two main types of LANs in use in the office environment: token ring and Ethernet. They both deliver the same result—high-speed interconnectivity. If greater speed or capacity is required, there are some very-high-speed LAN standards such as FDDI or 100-Mbps Ethernet. High-speed technologies are more expensive and are dealt with in more detail in Section 4.9.3. Table 4.1 gives a basic comparison between the technologies.

Ethernet and token-ring LANs have been formally adopted, with some minor changes, as the IEEE 802.3 and IEEE 802.5 standards; however, they are generally referred to by the original names.

Table 4.1
Comparison of LAN Types

Characteristic	Ethernet	Token Ring	FDDI	Fast Ethernet
Standard specification	IEEE 802.3	IEEE 802.5	ANSI X3T9.5	IEEE 802.3
Wire speed	10 Mbps	4 or 16 Mbps	100 Mbps	100 Mbps
Developed by	Xerox	IBM	ANSI	IEEE
Cable used	Coaxial or un-shielded twisted-pair (UTP) cable	STP	Optic fiber	UTP

Ethernet is much more popular than token ring. This is mainly because Ethernet is generally less expensive, and 10-Mbps Ethernet is almost as effective as 16-Mbps token ring. One result of Ethernet's popularity is that new products tend to be developed sooner for Ethernet than for token ring.

Token-ring LANs are well suited for connection to an IBM mainframe computer, as they were both developed by IBM; however, many products will equally well connect an Ethernet LAN to an IBM mainframe.

Token ring and a related LAN type known as *token bus* are better suited to real-time applications, such as computer-aided manufacturing. This is because the data transfer is deterministic (i.e., it is possible to determine the maximum delay that data will encounter). This advantage, however, does not add up to very much in an office situation.

It is *not* possible to mix token ring and Ethernet on the one cable because they use different protocols and speeds. If, for some reason, the two technologies must be integrated, you will require a 3-to-5 bridge to link an 802.3 LAN to an 802.5 LAN. Apart from the expense, the 3-to-5 bridge must convert from one protocol to another, and for this reason it is noticeably slower than bridges that link LANs of the same type. In most cases, organizations prefer to choose one technology and stick with it. Figure 4.2 illustrates token-ring and Ethernet LANs.

4.3.2 Radio LANs

Most LANs are interconnected using cable. It is also possible to use radio as the transmission medium. The hardware required is more expensive, but installa-

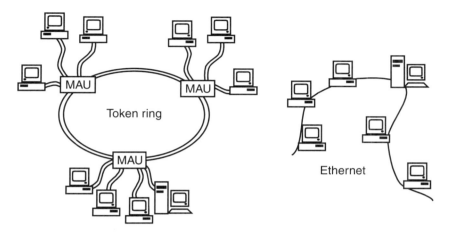

Figure 4.2 Token-ring and Ethernet LANs. In token-ring LANs, the "ring" is continued inside each *media access unit* (MAU).

tion costs are often lower, making it cost effective in temporary offices or in situations where frequent floor plan changes are envisaged. They are also useful where staff are constantly moving from place to place in the building with their portable computer (e.g., in a hospital). Finally, radio LANs can be mixed with traditional cabled LANs.

Many radio LAN technologies are limited to speeds of 1 or 2 Mbps—it pays to check what you are getting. It should also be noted that higher speed technologies usually have a smaller range, requiring more antenna (read dollars) for a given coverage area. Radio LANs create an additional risk as far as security is concerned because it would be feasible (though not trivial) for a hacker to log in to the network from outside the building. Encryption of the radio signals and good access control are required to prevent this. A good description of radio LAN technologies is available at [1].

4.4 Parts of a LAN

4.4.1 Cabling

The heart of any LAN is its cabling system. This is what ultimately carries the information between devices. It would be more accurate to refer to *transmission media* rather than cable, as wireless LANs use radio or infrared light rather than cable to carry the data. The different types of media used in LANs are summarized in Table 4.2, and popular cable types are illustrated in Figure 4.3.

Table 4.2
Different Types of Transmission Media Used in Local Area Networks

Medium	Typical LAN	Typical Span	Typical Speeds	Main Advantage
Thick coaxial cable (copper)	Ethernet	500m	10 Mbps	Noise immunity
Thin coaxial cable (copper)	Ethernet	185m	10 Mbps	Noise immunity and low price
STP (copper)	Token ring	100m	4, 16, and 100 Mbps	Ease of installation and noise immunity
UTP (copper)	Ethernet and token ring	100m	10 and 100 Mbps	Ease of installation and low price
Optic fiber	FDDI and Gigabit Ethernet	2 km	10, 100, and 1,000 Mbps	Noise immunity and high bandwidth
Radio	Ethernet	70–300m	2 and 10 Mbps	Ease of movement of workstations

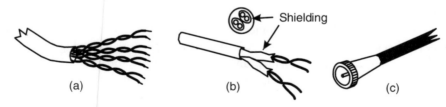

Figure 4.3　Some of the more popular cable types used in LANs: (a) UTP, (b) STP, and (c) coaxial.

In the past, the type of cable and the cabling topology (the way the cable interconnects the devices—see Section 2.3.4) was dictated by the type of LAN (Table 4.3).

There is a trend these days towards implementing a star topology from central wiring hubs, using high-grade twisted-pair cable (Figure 4.4), regardless of whether the LAN is token ring or Ethernet.[1] The two principal types

---

1. Token-ring LANs usually have a "logical" ring topology but a "physical" star topology. The ring is implemented within the hub. This is illustrated in Figure 4.2.

Table 4.3
Topologies of the Different LAN Types

LAN Type	Cable	Topology
Ethernet	Coaxial	Bus
Token ring	STP	Ring
FDDI	Optic Fiber	Ring

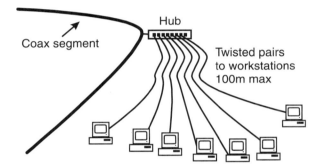

Figure 4.4 Twisted-pair cabling is always installed in a star configuration.

of twisted-pair cables, UTP and STP cable, must be manufactured to a specific standard. Both STP and UTP are capable of supporting 100 Mbps over a 100-m span, if installed correctly.

Using twisted-pair cabling in a star topology is part of the *structured cabling* approach. It can be used for both Ethernet and token-ring networks. It is usually a little more expensive than coax cabling (because there is additional cabling), but there are advantages: (1) cable failures tend to affect only one PC and not the entire network, and (2) it leaves you with wiring in place that should be suitable for upgrading to 100 Mbps if the need arises.

This second advantage is of great significance when installing a LAN in an old building where cable installation tends to be expensive and disruptive. The idea is that you can install a cabling system that will cater to your networking needs for many years to come without going to the expense of installing an all-fiber network. It is expected that, in time, all-fiber networks will drop in price to such an extent that they will be the preferred option for new installations.

4.4.2 Structured Cabling

Structured cabling is a system of prewiring a building for all LAN and telephone needs, both present and future. It makes extensive use of *patch panels,* where cable alterations can be made from a central location. Structured cabling is recommended for LANs with more than 40 or 50 workstations and can be used on both token-ring and Ethernet LANs.

It is important to note that the installation of twisted-pair cables must be carried out in accordance with strict guidelines; otherwise, they will not be suitable for 100 Mbps. Typical mistakes include:

- Untwisting the pairs excessively when terminating the cable;
- Putting too sharp a bend in the cable;
- Installing UTP too close to power cables, transformers, electric motors, or fluorescent lights [2];
- Breaking the shielding or faulty earthing in the case of STP.

In all cabling projects, it is necessary to comply with fire codes. Cabling poses two risks in the event of fire. First, the wrong type of insulation will cause the cable to assist a fire spreading throughout the building. Second, the insulation may emit toxic fumes when it is exposed to intense heat. The most stringent fire codes apply to cables installed in the *plenum* area. A plenum is the space above a false ceiling or below a false floor used by the air conditioning system to circulate the air. So-called plenum grade cable is considerably more expensive than standard cable.

4.4.3 Repeaters

Every cable type has a distance limitation caused by *attenuation.* Attenuation means the reduction in a signal's strength as you move further from the source of the signal. Attenuation normally gets worse as you increase the data rate (bits per second) of the signal.

Ethernet cabling has a naming convention that refers in part to the distance limitations of the cable. The various cabling options are referred to in Table 4.4.

For example, with 10Base5: the 10 refers to 10 Mbps, the Base refers to the baseband signal and doesn't concern us greatly, and the 5 refers to the 500-m distance limitation.

To overcome attenuation, some sort of signal booster is used. This device is called a repeater. It does more than simply boost the signal, however; it *regenerates* it. Regeneration involves reading the data on the input side and making a

Table 4.4
Names and Lengths of Popular Ethernet Cabling Options

Name	Alternative Names	Maximum Cable Segment Length
Thick coax	10Base5, thick net	500m
Thin coax	10Base2, thin net	185m
Twisted pair	10BaseT	100m
	100BaseT	
Optic fiber	10BaseF etc.	2 km (for 10BaseF)

fresh copy on the output side. On Ethernet networks the repeaters are normally *multiport* repeaters, thus providing a sort of junction box from which your cabling can spread out in various directions if needs be. The hubs used in structured cabling are usually multiport repeaters. Figure 4.5 illustrates cabling options for Ethernet LANs.

4.4.4 Workstations

On most LANs, the workstations are PCs. However, some applications require more processing power, and small minicomputers, such as Sun Sparc Stations, HP9000s, or DEC AlphaStations, are used as workstations.

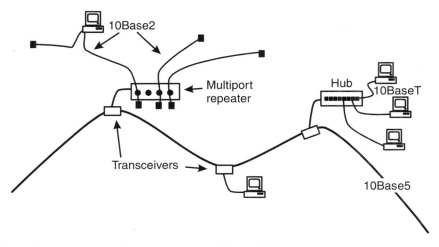

Figure 4.5 Some of the cabling options used on Ethernet LANs.

4.4.5 Network Interface Cards

Workstations and servers require the addition of a *network interface card* (NIC) to enable them to communicate on the network (Figure 4.6). Interface cards take up one expansion slot in the PC. New PCs can normally be purchased with these cards preinstalled.

The type of card purchased will depend on the type of network (token ring, Ethernet, or FDDI), and the type of connector on the card will depend on the transmission medium in use (Table 4.5). Some cards have more than one connector, as illustrated in Figure 4.6. This allows a workstation to be easily moved from one cable type to another (though it does not allow simultaneous connection to two LANs).

4.4.6 Terminal Server

Dumb terminals consist of little more than a keyboard and a screen. They are used as input/output devices on mainframe and minicomputers. They can be connected to their host computer via a terminal server, which is connected to the network via a network interface card. A typical terminal server will serve 16 dumb terminals, connecting them to any mainframe or minicomputer attached to the LAN (Figure 4.7). Terminal servers are also used for dial-in access to a LAN. In this case they are also referred to as *access servers*.

4.4.7 Servers

The most important servers on a network are those that store your shared information. PC networks generally have a simple PC-based file server. This will

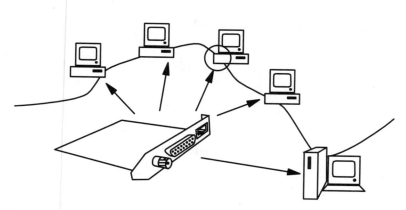

Figure 4.6 Each computer on the network needs a network interface card.

Table 4.5
Connectors Used With Various Cable Types

Cable	Typical Connector	Diagram
UTP	RJ45	
STP	IBM style data connector or shielded RJ45	
Thin coax	BNC	
Thick coax	15 pin "D" connector at PC and transceiver at coax	

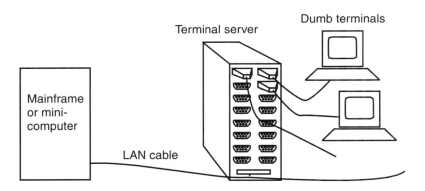

Figure 4.7 A terminal server is used to connect dumb terminals to a mainframe or minicomputer.

have a large disk storage area for all of the shared information and software. Other servers, such as minicomputers running UNIX, can both store and process information, leaving less work for the workstations.

Some networks do not have any servers for shared information. Instead, the workstations share storage resources with each other. These networks are called *peer to peer* and are discussed in Section 4.6.

With a fax server, a document is prepared on one of the workstations and then submitted to the fax server for transmission over the telephone network. The fax server looks after the queuing and dialing operations, leaving the workstation free to get on with other tasks.

Fax servers can also receive fax messages but this is not as common because of the difficulties involved in distributing the faxes to the correct recipients. One of the most effective methods of distributing incoming faxes is to use *direct dial in* (DDI; DID in the U.S.). DDI has already been mentioned in Section 3.4.3 as a means of directly dialing the extensions in a PBX from the public telephone network. In the case of the fax server, you must rent an additional DDI number for each fax user and program the fax server to distribute the incoming fax to the user's PC based on the user's DDI fax number. To get a DDI service from the telephone company, you will more than likely have to rent an ISDN line.

CD-ROM servers can give many users concurrent access to CD-ROM databases. The server will normally contain a number of CD-ROM drives to allow a number of CDs to be online at the same time.

4.5 Software

4.5.1 Network Operating System

The *network operating system* (NOS) running on a server carries out the following functions:

- File sharing;
- File and record locking;
- Security (i.e., passwords);
- Print spooling (temporary storage of print information before it gets to the printer).

The operating system on the server will influence the performance of the server, while the operating system running at the workstation (DOS, Windows, Mac OS, UNIX) will influence what software can be run at that particular workstation.

Currently available NOSs for servers include:

- Novell NetWare;
- UNIX;

- Microsoft Windows NT;
- Microsoft LAN Manager;
- Banyan Vines;
- IBM OS/2 LAN Server;
- Apple AppleShare.

Novell NetWare, with its many versions, claimed 67% of the market for PC-based NOSs in 1995 [3]. It can be installed on an Intel-based PC, which then becomes a file server. Its popularity is partly because it is supported by so many end-user applications and has a memory-efficient *requester* (see Section 4.5.2). Windows NT has gained market share in recent years particularly on smaller networks.

File server–based networks became very popular in the late 1980s because it was so easy to convert PC software into a network version. In many cases it meant little more than ensuring that the software marked a file as read-only whenever it was in use. As far as the software was concerned, the file server was just another disk drive.

The UNIX operating system is normally used on minicomputers. It is traditionally used in host-terminal mode, but more recently it is used in a client/server environments. Unlike a file server, which is only responsible for the storage of data, a UNIX server is responsible for both storage and some or all of the processing of that data. Hence, it generally requires more processing power than a simple file server.

4.5.2 The LAN Requester

The LAN requester is software that must run on each workstation attached to the LAN. When an application makes a request to its host operating system to access a resource, the host operating system passes the request on to the requester. This will then decide if the resource is local or remote. If the resource is remote, the requester will decide at which server the resource is held and send a request to the server software to run an application or fetch a file.

The requester may be specific to the protocol in use on the server. So, for example, a server running Novell NetWare normally requires you to run a protocol called IPX/SPX, while a UNIX server on the same network may require you to run the *transmission control protocol/Internet protocol* (TCP/IP). Thus, you may require more than one requester running in memory. Alternatively, the requester may have more than one protocol built in—the latest versions of requester often do. One of those protocols will invariably be TCP/IP, given the popularity of this protocol today.

4.5.3 Application Software

Application software refers to the end-user software (e.g., the word processor, spreadsheet, email, or database software used by the network users). In general a network version of this software will be installed on the file server. Users then load whichever program they wish to use onto their workstation. They can then use the program to access data on the file server or on their local disk.

An important point to remember about application software stored on a file server is that it is the workstation that runs the software and not the file server. This has two important implications. First, processing speed will be determined to a large extent by the speed of the workstation. Second, if large quantities of data must be processed (e.g., a database search), all of the data must pass through the workstation for processing. If the data source is the file server, then all this data must be transferred across the network. Client/server computing can improve performance significantly in this area.

4.5.4 The Development of Client/Server Computing

Network architecture has progressed in the following manner:

$$\text{Host terminal} \rightarrow \text{File servers} \rightarrow \text{Client server}$$

The classic example of the host-terminal environment is the mainframe computer being accessed by a large number of dumb terminals. Information flow consists of keystrokes towards the mainframe (or minicomputer) and screen information back towards the terminal. This does not create a very large data flow but leaves all the processing to one machine.

With the advent of PCs, processing power is now available in the user's workstation. Many PC networks were, and still are, based on file servers. The classic example of this is a network based on Novell's NetWare operating system. File servers are simply a common storage area; they do very little to enhance the performance of a slow workstation. The workstation still has to do all of the processing. Data flow on this type of network consists of the transfer of the file contents to the workstations for processing and back to the server for storage. This results in large data transfers. Very often, only a small portion of the information transferred is actually needed. Figure 4.8 illustrates data flow on a file server–based database.

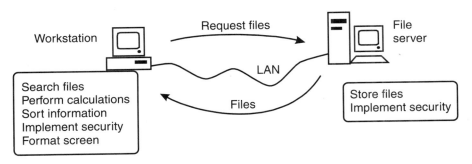

Figure 4.8 Data flow on a file server–based database.

In the client/server model, the data processing is shared between the workstation (client) and the server. So, for example, in a database application the workstation would be responsible for providing the menus, buttons, and dialogue boxes to the end user, and the server would be responsible for searches and perhaps sorting of data. The workstation is also responsible for formatting the data as it arrives from the server and perhaps for doing further calculations on the returned data. A well-designed client/server database application will generate far less network traffic than a simple file-based database. Figure 4.9 illustrates a client/server database.

Client/server applications have two parts to their software: the client software (front end) running on the workstation and the server software (back end).

Most network operating systems (including Novell NetWare) include support for client/server applications in their latest versions. Thus, a single server will often double as a file server and an application server.

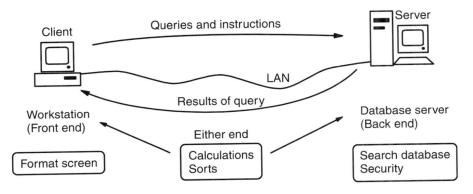

Figure 4.9 Client/server database.

4.6 Hierarchical Versus Peer to Peer

Many large networks are hierarchical, with the file server at the top of the hierarchy and the workstations below. This is the case with the most popular PC-based system, Novell NetWare. All information flow depends on the file server. This means that if information is to be shared, it must be stored on the server. Even user-to-user email is sent via the file server.

It is, however, possible to set up a small peer-to-peer network without a file server. Such a network allows the users to share files on one another's hard disks and to share resources, such as printers. Peer-to-peer networks can make great sense where there are only a few users (three or four), especially if they are familiar with their PCs. Apart from not having to invest in a file server, the networking software will be less costly. The most popular peer-to-peer software includes LANtastic, Windows for Workgroups, Windows 95, and Windows NT Workstation. It is important, however, to realize that peer-to-peer networks are not as easy to manage and require cooperation, trust, and know-how among all the users.

It is also possible to combine the two concepts by using peer-to-peer networking alongside hierarchical networking. By this means, small workgroups can be established, within which people have access to one another's hard disks, while at the same time everyone has access to the servers on the network. A few words of caution about this approach are in order, however. First, each PC on a peer-to-peer network acts in a similar way to a server, regularly broadcasting packets that advertise its presence on the network. This increases network traffic and reduces the performance to the user of that PC. Second, many peer-to-peer NOSs use protocols that cannot be routed, thus making them unsuitable for large enterprise networks (see Sections 12.5.3 and 12.5.4).

4.7 Network Management

The usefulness of a network to a business will depend very much on the way it is managed. The most important thing is to make life as simple as possible for end users. Technology alone will not achieve this. A good network manager who understands both the technology and the needs of the user is required. Good management tools will in turn make the network manager's job a lot easier.

4.7.1 The Functions of the Network Manager

Many businesses that switch from mainframe computers to LANs are surprised at the hidden costs of network and PC management. Managing a medium-sized network of 50 users can represent a full-time job for one person. This does not include the often significant amount of time spent by users attempting to solve their own problems. Among the tasks involved are:

- Cable management;
- Installing and configuring new PCs;
- Installing and configuring new software and software upgrades both on the server and the PCs;
- Configuring and reconfiguring user rights;
- Implementing data backups and security;
- Monitoring utilization of bandwidth and storage and taking action before problems occur;
- Problem solving (e.g., "I cannot print," "I cannot run this application");
- Attending training courses to ensure that all the above work can be carried out.

Much time is spent on end-user support, which has prompted the computer industry to develop *network computers* (NCs). The idea is to develop an end-user computer that is less configurable than a PC and thus less costly to manage.

4.7.2 Simple Network Management Protocol

Most network hardware can include remote management capabilities. This means that they can be controlled from a workstation elsewhere on the network. For example, a device like a router in a branch office can be configured from a central management center, and alarms can be delivered to the management center (provided a link remains to carry the information). *Simple network management protocol* (SNMP) is a standardized method of allowing devices from a variety of suppliers to be managed from a single workstation. NOS providers may, however, offer network management tools not based on SNMP. Making use of such a system may limit your choices when it comes to the purchase of additional equipment for the network.

4.8 Shared Resources

When considering shared resources, the following potential drawbacks must be kept in mind: cost, queuing, complexity, and increased network traffic. For example, a fax server consists of either a PC or a customized box, with a number of fax-modems installed, plus a significant multiuser license fee for the software. In a small organization, it could prove to be more economical and simpler to install a separate fax-modem in each PC that needs to send faxes.

Queuing for resources results in delays. For example, if one person sends a fax message to a mailing list of 100 fax numbers, everyone else who wants to send a fax will experience a delay of a number of hours. This situation can be alleviated by scheduling nonurgent faxes to be sent at night or by giving a priority to each fax message.

It is important not to put all your eggs in one basket. For example, it is more prudent to install two medium-priced printers than one very expensive printer. If one printer develops a fault, you still have the second one.

4.9 Increasing LAN Capacity

The wire speed of a LAN is ridiculously fast when considered in the context of two PCs connected to one another. What you must remember is that this speed, often referred to as bandwidth, must often be shared between *all* the devices on the network. Only one device is allowed to transmit on the cable at one time; hence, if two devices have data for transmission, one must wait until the other has finished. This wait will be very short because of the high transmission speed (e.g., 10 Mbps) and because there is a limit on the amount of information any device can transmit at a time.

But what happens as you increase the number of devices that have data for transmission? At some stage your network will become congested. Mild congestion will cause the network to slow down. Heavy congestion can result in total network failure. Congestion is rarely a problem on LANs with fewer than 30 PCs, but as the number of PCs increases so does the risk of overload.

The number of PCs that can be connected to a single LAN depends on how much network traffic each PC generates. For example, a PC being used to look up a central database of scanned images will generate a lot more traffic than one used for simple word processing.

If a LAN becomes overloaded, you have a number of options. First, you can check if the network bandwidth can be used more efficiently (e.g., by storing more information or software on the local workstations).

4.9.1 Local Bridges

If fine tuning the applications doesn't work, you could split the network in two, linking the two LANs with a device called a *local bridge*. A local bridge will only pass information to the other side if it has to. A server will be required on both sides of the bridge; otherwise, too much traffic will have to cross the bridge. If everything is done correctly, the traffic on each LAN will be almost half what it used to be.

This approach can be taken a stage further by breaking the network into a number of smaller networks, each connected via a bridge to a backbone network. The backbone could be a higher speed LAN if necessary (e.g., an FDDI backbone running at 100 Mbps). The servers would be directly connected to the backbone.

4.9.2 Switching

If distance is not an issue then the high-speed backbone can be collapsed into a box called a switching hub or simply a switch. These collapsed LAN backbones can run at even higher speeds than can the FDDI backbone. The servers can be connected to the switch using a 100- or 1,000-Mbps technology, while the other LANs are connected at 10 Mbps (or 100 Mbps if required). Figure 4.10 shows a typical switched Ethernet LAN.

4.9.3 High-Speed LANs

A third option, upgrading the entire LAN to a high-speed LAN operating at 100 Mbps, also exists. This is the most expensive option and will involve the replacement of all the network cards, repeaters or hubs, as well as possibly replacing cabling if suitable twisted-pair cable is not in place. There are three technologies which work at 100 Mbps:

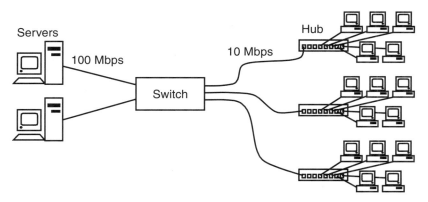

Figure 4.10 Using a switch to increase the capacity of a LAN.

- Fiber-distributed data interchange–copper-distributed data interchange (FDDI-CDDI);
- 100BaseT (fast Ethernet);
- 100VG-AnyLAN.

FDDI

FDDI is the oldest 100-Mbps technology. It was originally specified for use on an all-fiber network. CDDI was later developed to give 100-Mbps connections to end users over copper cable at a more reasonable cost. FDDI and CDDI are quite expensive compared to the other two technologies. FDDI is, however, a very mature standard, and the fiber runs can be up to 40 km if necessary.

100BaseT (Fast Ethernet)

The 100BaseT products emerged in 1995, and by 1997 the prices were close to their 10-Mbps counterparts. Part of the marketing strategy is to produce cards that support both 10 and 100 Mbps. This allows a network manager who is considering an upgrade in the near future to start installing high-speed cards straight away—particularly in new workstations. There are three main versions of 100BaseT—100BaseTX, 100BaseT4, 100BaseFX—each suitable for a different cable type [4].

100BaseTX works on two-pair copper cable. High-quality cable is required—either category 5 UTP or category 1 STP. 100BaseT4 requires four pairs but can work on lower quality cable (category 3, 4, or 5 UTP). 100BaseFX works over two optic-fiber strands.

100VG-AnyLAN

This technology emerged about the same time as 100BaseT [5]. Early indications are that it is not nearly as popular, despite similar pricing and some good features. It is designed to work over *voice grade* (VG) cable—category 3, 4, or 5 UTP. Like token ring, it is deterministic and traffic can be prioritized. These features are likely to be of interest in a manufacturing environment. This technology is also likely to give higher quality video conferencing across the LAN than fast Ethernet does.

Distance Limitations at 100 Mbps

A traditional 10-Mbps Ethernet LAN can have a diameter of up to 3 km before you will be forced to spend money on a bridge, router, or switch. This is not the case with 100-Mbps LAN connections. Thus, 100-Mbps Ethernet LANs tend to use switches every 205m when using copper and every 405m when using fiber. This adds to the expense of 100-Mbps Ethernet. A similar situation exists for 100VG-AnyLAN.

Electromagnetic Interference

One possible snag involved in upgrading to 100 Mbps on copper is that after the upgrade, the network might be over the limit in terms of the level of *electromagnetic interference* (EMI) it emits. EMI can interfere with the correct operation of other electronic equipment, so legal limits are set. In Europe, the legal limit on EMI is lower than in the United States (i.e., it is harder to comply). Approaches to reducing EMI include:

- Using STP or fiber;
- Ensuring quality workmanship on UTP cabling;
- On UTP schemes, using network cards and hubs that utilize all four pairs (100BaseT4 or 100VG-AnyLAN).

It is interesting to note that 90% of German LANs use STP cabling or some shielded variant of four-pair cable [6]. This is largely due to the German authorities reserving the right to close down a network that emits too much EMI.

Gigabit-per-Second LANs

At the time of writing, work was still ongoing on a specification for an Ethernet standard for a LAN operating at speeds up to 1,000 Mbps (1 Gbps). The initial work specified fiber, but a copper-based standard is also due. The standard is due to be ratified early in 1998; however, working products have already been demonstrated in advance of a formal standard. Very few, if any, end users will require such high bandwidth at their workstations. The main focus of this technology will be in providing high-bandwidth backbone links between switches and between servers and switches [7].

References

[1]   Rune, Torben, "Wireless Local Area Networks," http://www.netplan.dk/wireless.htm (no date but appears to be 1995 or 1996).

[2]   Macaulay, Peter, "Cabling FAQ," comp.dcom.cabling or http://www.cis.ohio-state.edu/ hypertext/faq/usenet/LANs/cabling-faq/faq.html, 1995.

[3]   "Novell NetWare Network Operating System—Product Report," *DataPro CD-ROM,* Sept. 1996.

[4]   3Com, "100BASE-T Fast Ethernet: A High-Speed Technology for Cost-Effective Scaling of 10BASE-T Networks," http://www.3com.com/0files/strategy/fasteth.html, June 1996.

[5]   Russel, Richard G., "Richard's Unofficial 100VG AnyLan Web FAQ," http://www.io.com/~richardr/vg/index.html, Jan 1997.

[6]   Greenfield, David, "Wire act leaves LANs dangling," *Datacommunications International,* February 1996.

[7]   "Gigabit Ethernet Overview—May 1997," Gigabit Ethernet Alliance, http://www.gigabit-ethernet.org, 1997.

Further Reading

Hunter, Philip, *Local Area Networks Making the Right Choices,* Wokingham, England: Addison-Wesley, 1993.

5

The Telephone Network

5.1 Introduction

The ordinary telephone network, or *public switched telephone network* (PSTN), has been described as the largest machine ever built. It is truly global, spanning virtually every country on every continent, with about three-quarters of a billion telephone lines connected to the network [1]. The PSTN is over 100 years old, and as it has developed each new innovation has to be compatible with the existing components of the network. For example, modern digital exchanges with sophisticated control architectures and signaling systems have to interwork with older electromechanical exchanges using older, simpler signaling systems.

The ITU-T (formerly the CCITT) has the responsibility of recommending the standards to be used throughout the worldwide PSTN and has achieved this task more or less successfully. However, incompatibilities can and do exist.

The PSTN is designed for speech. Users need to be able to understand what is spoken and, to a lesser extent, to be able to recognize the speaker's voice. Although the human ear is capable of detecting a large range of frequencies, typically 10 to 15,000 Hz, it is only necessary to transmit a small part of this frequency range to meet this need. ITU-T recommendations state that telephone circuits must transmit frequencies in the range from 300 to 3,400 Hz. This is called the *bandwidth* and is one of the main constraints on using the PSTN for nonvoice applications.

The PSTN can be used to transmit digital signals such as those transmitted by computers, facsimile machines, and video telephones. However, when using the traditional PSTN, these digital signals must be processed (by modems) so that they have the same characteristics as the human voice and in particular conform to the bandwidth requirement. Recent developments of the PSTN, in particular ISDN (which is covered in the next chapter), can enable direct digital access, removing these constraints.

The PSTN comprises the following main components (Figure 5.1):

- *Transmission:* a system of digital transmission circuits interconnecting the exchanges;
- *Switching:* the function of the exchanges (also called *central offices*);
- *The local access network:* the network of cables that connects the telephones to their local exchange.

5.2 Transmission

The transmission network is a high-capacity transport network connecting the switching centers (or exchanges). It is normally semimeshed, providing a high degree of reliability between the main centers. Although originally *analog,* the transmission network was the first part of the PSTN to be converted to *digital.* In most developed countries, the transmission network is now completely digital. Modern digital transmission systems have the following features:

- *High reliability.* The equipment is normally very reliable. In addition to this, the network operators view transmission failures as very serious. Thus, transmission failures are attended to immediately. Due to the transmission network being semimeshed, signals can be sent by alternative routes in the event of failures.
- *"Perfect" quality.* Once a signal is converted to digital, the digital signal can be reproduced identically at any stage. Thus, digital transmission is

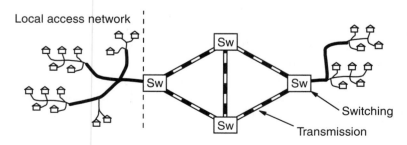

Figure 5.1 Components of the PSTN.

distance independent. If digital transmission is used for the entire connection, the speech quality on an international call between two continents is identical to that on a local call.

Digital transmission relies on two concepts: *pulse code modulation* (PCM), which is a method used to convert from analog to digital (see Section 2.2.2), and *time-division multiplexing* (TDM), which is the system used to transmit a number of digital signals over a single transmission medium.

5.2.1 Time-Division Multiplexing

Multiplexing refers to the transmission of a number of channels on a single shared transmission medium. A number of methods can be used, but in modern digital networks TDM is the most common.

TDM simply means that each channel is given exclusive access to the transmission medium at fixed periodic time intervals called *time slots*. A number of time slots combine to form a *frame*. The frame structures can vary, but the *E1* format (used in Europe) will be used to illustrate the principles of TDM. In the E1 format, each frame consists of 32 time slots.

To use TDM in the PSTN, the speech is first converted into a 64-Kbps digital signal using PCM. An E1 frame consists of 32 time slots, each carrying a 64-Kbps channel. The E1 frame must therefore be transmitted at 2.048 Mbps (32 × 64 Kbps), which is normally written as 2 Mbps for convenience. The frame repeats every 125 µs.

To illustrate the principle of TDM, consider two telephone conversations between A and B and between C and D, respectively (Figure 5.2).

The speech from A is converted using PCM into a 64-Kbps digital signal consisting of an 8-bit sample every 125 µs. The first sample from A is placed into the frame in time slot 1. Then, 125 µs later, the second sample is ready, just as time slot 1 appears in the next frame. As the samples and the frames both occur every 125 µs, subsequent samples from A will all be transmitted in time slot 1 of subsequent frames.

Similarly, all of the speech samples from C will be placed in time slot 2 of the frames.

At the far end, all speech samples in time slot 1 are extracted from the frames and used to reconstitute the 64-Kbps digital signal. These are decoded and filtered to reproduce the original analog speech signal from A, which is then transmitted to B.

Similarly, the samples in time slot 2 will be extracted, decoded, and filtered into the original analog speech signal from C and then transmitted to D.

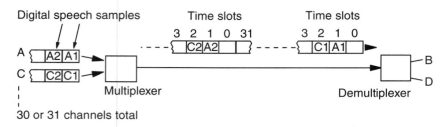

Figure 5.2 Principle of TDM using E1 system.

A complete E1 system in fact consists of two systems, one for transmission (A to B) and one for reception (B to A).

Not all of the time slots in an E1 system are available to be used for speech channels. Time slot 0 is always used by the system itself as a synchronization and error-control channel. In many systems, a further time slot is used as a signaling channel to set up and tear down calls and to activate special features. Thus, the E1 system is often referred to as *30-channel PCM*.

Another system in use is called the *T1* system or *24-channel PCM*. This system uses a frame containing 24 channels, all of which are normally available for speech (Figure 5.3). A variety of methods is used to carry the signaling, though most involve "stealing" the lowest order bit from some of the speech samples. This affects speech only very slightly but means that only seven of the eight bits in the sample can be used reliably for data. Therefore, 56 Kbps (7 bits × 8,000 per second) is the standard digital data rate where T1 systems are used. T1 systems are used mainly in the Unites States, Canada, and Japan.

5.2.2 European Transmission Systems

The E1 system forms the *first-order* or *primary-rate* PCM system in the European system. By combining four E1 systems, an 8-Mbps E2 or *second-order* system can be created. Similarly, second-order systems are multiplexed to create third-order systems. Table 5.1 gives the details up to the fourth order or *quaternary rate* system.

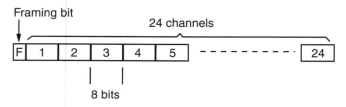

Figure 5.3 T1 (24-channel) PCM system.

Table 5.1
European Transmission Systems

System	Name	Bit Rate	Number of Channels
First order (primary)	E1	2 Mbps	30
Second order (secondary)	E2	8 Mbps	120
Third order (tertiary)	E3	34 Mbps	480
Fourth order (quaternary)	E4	140 Mbps	1,920

Note that bit rate is always stated as the approximate rates shown in Table 5.1. In fact, a 2-Mbps system has an actual bit rate of 2.048 Mbps (as mentioned previously), while an 8-Mbps system has an actual bit rate of 8.448 Mbps.

5.2.3 Other Transmission Systems

The transmission system shown in Table 5.1 is used in Europe and Australia. Separate standards were developed in the United States and Canada and in Japan. This must be kept in mind when implementing international leased lines. Table 5.2 outlines the four main transmission systems.

5.2.4 Synchronous Digital Hierarchy

The transmission systems outlined in the previous section are examples of *plesiochronous digital hierarchies* (PDH). The term *plesiochronous* refers to the transmission network not being synchronized to a single national clock. PDH systems are the main transmission systems currently in use throughout the world. To remove some difficulties associated with PDH, Bellcore (the research arm of the Bell operating companies in the U.S.) proposed a new synchronous hierarchy called *synchronous optical network* (SONET) in 1985. An American standard for SONET was issued in 1988. In 1988 the CCITT (now ITU-T) also issued recommendations for a similar system called *synchronous digital hierarchy* (SDH). Developments and revisions to these standards are ongoing. There is a high degree of compatibility between the American SONET and the European SDH, both using the same frame structure and bit rates.

The basic element of SDH is the 155-Mbps STM-1 frame. SONET uses a 51-Mbps signal called OC-1. An OC-3 signal is equivalent to an STM-1. A

Table 5.2
Transmission Systems

System	U.S. and Canada	Japan	Europe and Australia
Primary	1.5 Mbps (T1)	1.5 Mbps	2 Mbps (E1)
Secondary	6 Mbps (T2)	6 Mbps	8 Mbps (E2)
Tertiary	45 Mbps (T3)	32 Mbps	34 Mbps (E3)
Quaternary	–	98 Mbps	140 Mbps (E4)

flexible multiplexing hierarchy means that an STM-1 frame can carry, for example, any of the following European PDH signals:

- 1×140 Mbps;
- 3×34 Mbps;
- 63×2 Mbps;
- 1×34 Mbps plus 42×2 Mbps;
- 2×34 Mbps plus 21×2 Mbps.

The STM-1 frame can also carry American PDH signals, as well as ATM and FDDI.

With SDH and SONET it is much simpler to *drop and insert* one or more of the signals at intermediate points along a transmission path than with PDH.

Other important features of SDH and SONET are the inclusion of network management capabilities to automatically reconfigure the network in the event of failure and the inclusion of standard optical fiber interfaces.

The main advantages of SDH and SONET from the customer's point of view are:

- Faster response times from network operators to customer requests;
- Improved reliability;
- Higher quality.

5.2.5 Transmission Media

The most important transmission media in use in the PSTN are optic fibers in the backbone transmission networks and twisted-pair cables in the local access network. Microwave radio links are used to provide alternative transmission paths in areas where it provides a lower cost solution than fiber. Coaxial cable was once the main medium used for the backbones but has been superseded by fiber.

5.3 The Switching Network

5.3.1 Introduction

As mentioned in Section 2.3.3, switching is used to reduce the amount of communication channels required in a network. Circuit switching is used in the telephone network. Switching is performed to connect a local access line to a trunk circuit, to connect two local access lines, and to connect two trunk circuits. The point at which switching takes place is called an *exchange* or *central office*. On a typical long-distance telephone call, switching will take place at a number of intermediary points in *transit exchanges* as well as at the *local exchanges* to which the users' local access lines are connected. There are a number of different types of exchanges and switching techniques that can be used. The oldest type of exchanges were manually operated exchanges of the *magneto* and *central battery* types. These have been phased out in modern networks. Automatic exchanges range from the earliest *strowger* (or *step by step,* also phased out in modern networks) types to the *crossbar* to the modern *digital* exchanges. In modern networks, only digital exchanges are currently installed, although existing crossbar and even strowger exchanges are retained in some networks if they are working adequately.

5.3.2 Exchange Layout

The layout of an automatic telephone exchange is shown in Figure 5.4. The generic layout shown is applicable to crossbar, SPC-controlled crossbar, and digital exchanges.

Subscriber Switching Stage

The telephones are connected via access lines to a *subscriber switching stage.* Every line has its own individual circuit in the subscriber stage. When this circuit detects an *off-hook* signal indicating that a call is about to be initiated, the line is connected to a common device to receive the dialed digits and a dial tone is sent on the line. When sufficient digits have been dialed, the circuit can then be switched to an outgoing channel connected to the input of the *group switching* stage. In some cases, if the dialed digits indicate that the call is destined for a

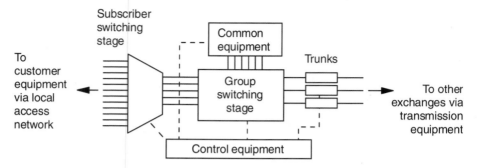

Figure 5.4 Layout of an exchange.

line connected to the same subscriber stage, that call can be fully switched within the subscriber switching stage.

The entire subscriber switching stage of an exchange is implemented in a number of separate groups. Typical sizes of groups would be 1,000 line groups (crossbar) or 2,048 line groups (digital). The only connection between the separate groups is via the group switching stage. Based on the assumption that only a percentage of lines will be active at any stage, each subscriber switching stage has a limited number of outgoing channels connecting to the group switching stage, typically between 60 and 500 channels for a digital group containing 2,048 lines. Thus, the subscriber stage performs *concentration* on outgoing calls, concentrating 2,048 lines into 500 inputs for the group switching stage, and *expansion* on incoming calls, expanding 500 outputs from the group switching stage into 2,048 lines. The degree of concentration/expansion varies—in a busy urban area, normally the maximum number of circuits is fitted, whereas in suburban and rural areas, often the minimum number of circuits is all that is required.

As only a percentage of inputs (i.e., access lines) can be connected to the outputs, this type of switching involves *blocking* (i.e., if all outputs are busy, the other inputs are blocked from the switch).

Remote Subscriber Unit. In digital exchanges, the subscriber stages also contains PCM circuitry to convert analog signals to and from the access line to digital. Thus, the connections to the group switching stage is via 2-Mbps (or 1.5-Mbps) digital links. As one of the features of digital transmission is distance-independence, the subscriber stage can be situated anywhere. In low-density areas, the subscriber switching stages can be placed close to groups of customers, thereby reducing the length (and cost) of lines in the access network. These subscriber switching stages are termed *remote subscriber units* (RSUs).

The main drawback is that the RSU is dependent on the digital links to the main exchange for the routing of any calls outside that RSU. Furthermore, some RSUs are also dependent on these links for control information and will be unable to connect even local calls if the links fail.

Group Switching Stage

The *group switching stage* of the exchange is the main switching point and is always located in the main exchange building. Group switching stages are *non-blocking* in that all inputs can be connected to all outputs. The connections to the group switching stage include:

- Circuits from the subscriber switching stages;
- Circuits from trunk devices (coming from the transmission network);
- Common devices such as signaling equipment announcement devices.

Trunk Devices

Trunk devices interface to circuits connecting to other exchanges, normally via the transmission network. There are three categories of trunk devices: *outgoing,* which control circuits for calls destined to other exchanges; *incoming,* which control circuits for calls coming in from other exchanges; and *bothway,* which can control calls in both directions. In modern networks, the trunk devices terminate E1 or T1 transmission links (2 or 1.5 Mbps).

Common Devices

Common devices are not tied to a particular user or a particular transmission circuit but are held in a common pool. The first group of common devices includes equipment that is used during the setup or clear down of a call. They are used for only the short time that a particular call connection is being established and then released (to be used in the setup or clear down for other call attempts) so a relatively small number of devices can handle a large number of calls. Examples of this would be signaling senders and signaling receivers. The other group of common devices are used for particular types of calls, such as test calls, diverted calls, and conference calls. Examples would be announcement machines, devices for three-party calls, and test equipment. Because only a small percentage of calls would require these facilities, only a relatively small number of these devices are required.

Control

Control equipment is not used in the call path, but is used to *control* the setup of the switching equipment in both the subscriber-switching and group-switching stages. In general, the degree of sophistication of the control equipment determines the amount of enhanced services a network can offer.

5.3.3 Analog Exchanges

In analog exchanges, switching is performed on the original analog signal. A physical connection is made at the switching point between the incoming and outgoing circuits. This is done by electromechanical means. If connecting to a digital transmission network, PCM is performed after the switching stage (for outgoing calls).

Crossbar

As already mentioned, the oldest analog exchanges (step by step) have been phased out in most developed-world networks. Crossbar exchanges are still in use but have been phased out in some networks.

Operation.   A crossbar switch consists of a matrix of horizontal and vertical levers, with a set of contacts at each of the intersections. The switches are set by means of electromagnets.

The switches are controlled by devices called *markers.* A marker is a logic device with the processing being carried out by electromechanical switches called *relays.* The data for the marker is stored by hard-wired connections.

Features.   The major advantage of crossbar over step by step is the separation of the control function from the switching function. This means that changes are easier to implement, although they still require major rewiring of the logic in the markers. Although pulse dialing was initially used, it is not an built-in part of the switching mechanism. Thus, many types of crossbar exchanges can be upgraded to accept tone dialing. The common control structure means enhanced features such as detailed billing can also be implemented by adding modern control units.

SPC-Controlled Crossbar

Operation.   SPC stands for *stored program control.* In effect, the common control functions of a crossbar exchange are carried out by *processors,* with the instructions being *stored* as a *program* in memory. The term predates the widespread use of computers; we would probably call it a *computerized* crossbar exchange nowadays.

The switching mechanism is the same as for crossbar, so although *digital* control is used, this type of exchange is still an *analog* exchange.

Features. The advantage of SPC is that it allows changes in the network to be made more quickly and easily. In particular, it makes the provision of new enhanced services possible.

5.3.4 Digital Exchanges

Digital Switching

The term *digital switching* refers to the idea that the switching is performed on the 64-Kbps digital bitstream produced *after* the PCM process. The two ways in which a digital bitstream may be switched are called *time switching* and *space switching*. Both switching methods are used in most digital exchanges, although some exchanges use only time switching. The objective in digital switching is to be able to switch any 64-Kbps channel from any time slot in one 2-Mbps system to any other time slot in any other 2-Mbps system.

Control Structures in Digital Exchanges

There are two common types of control structures used in digital exchanges. One type is termed *central control,* in which a single main processor controls all the parts of the entire exchange. The other type is called *distributed control,* whereby control is distributed among a number of processors. In distributed control, each processor controls only part of the exchange—for example, a group of trunk devices, the group switching stage, or a subscriber switching group. There are certain advantages and disadvantages associated with each type.

Central Control. In this case, a single *central processor* controls the entire exchange. The central processor is usually duplicated or triplicated to avoid the possibility of failure, as failure of the central processor means that the entire exchange ceases to operate. In this case, the central processors are said to be a *central processor group.*

In the duplicated central processor type, one processor is nominated as *executive* and therefore has control over the exchange. The other processor is nominated as *standby* and follows the execution of the executive exactly. It executes the same software and reads the same data from the exchange as the executive processor, but does not have any control over the exchange. In the event of a failure in the executive processor, the standby processor can take over immediately without any calls being lost.

In the triplicated processor type, all three central processors work identically. In the event of a failure in one processor, the exchange obeys the instructions from the other two on a *majority* principle.

Obviously, a failure of the central processor causes the most serious alarm to be raised in the exchange and must be repaired by the maintenance staff as a priority.

The exchange equipment is not directly controlled by the central processor, but by *regional* and/or *device processors,* as direct control of every device would place an impossible load on the central processor. These regional and device processors take care of routine tasks, such as scanning access lines for on-hook or off-hook conditions or detecting if a digit has been dialed, but do not make any major decisions. For example, if a regional processor detects that a particular digit has been dialed, it sends a signal to the central processor. The central processor will analyze the digit in conjunction with the previous dialed digits and will decide, for example, that the call should be switched to a particular trunk device. It then sends signals to the regional processors controlling the switching stages to switch the call to that particular trunk device.

The advantage of central control is *flexibility.* It is easier to make major changes to the operation of the exchange, as often only the software in the central processor needs to be updated. In particular, it is easier to provide new enhanced services such as virtual private networks, closed user groups, and intelligent networking services. From the exchange manufacturer's point of view, it is also easier to modify the design of the exchange. The same basic design of exchange can be used as local, trunk, international, mobile, or ISDN by changing some of the hardware and simply changing the software in the central processor. This should translate into an advantage for telephone companies and users in the form of reduced development costs being passed on.

The disadvantage of the common control type exchange is the dependence on the one central processor group. Duplication (or triplication) means that a *hardware* fault in one processor should not be a problem, but multiple faults can occur. More importantly, *software faults* can (and do) occur. As both sides execute the same software and contain the same data, both sides of a duplicated central processor will fail, resulting in the entire loss of service in the exchange. Usually the software can be restarted in a minute or two, but it is not unheard of for an entire exchange to be out of service for a few hours!

Software obviously cannot *go* faulty, so how do software faults occur? Quite simply, the software has been flawed since the beginning. What has happened in a *software fault* situation is that the particular block of software containing the fault has been executed for the first time, with the particular set of data that cannot be handled by the software. For example, a customer has activated a new enhanced service under conditions that had not been tried before. By resetting the central processor and breaking down the call that activated the faulty software, the software fault can be temporarily remedied.

The amount of time for which an exchange is out of service is called *downtime*. Downtime is the subject of major investigations within telephone companies and between the telephone company and the exchange manufacturer. The objective is obviously to reduce, and if possible eliminate, downtime.

Distributed Control. In distributed control, the exchange equipment in each part of the exchange is controlled by its own individual processor. No individual processor has overall control of the exchange. Each processor makes its own decisions regarding the equipment it controls. As a call is processed through the exchange, the control of the call is passed from stage to stage by communication links between the individual processors.

The advantage of distributed processing is that there is no single point of failure for the entire exchange. If a processor fails due to a hardware or software fault, only the part of the exchange controlled by that processor is out of service. Thus, if a processor controlling a group of trunk devices fails, service will be degraded but not completely disrupted.

The disadvantage of distributed processing is that the introduction of changes to the exchange, such as the introduction of a new service, is more difficult than with a centrally controlled exchange. This is because often a number of different processors may need to be changed to cater for the new service. Additionally, the signaling on the communications links between the processors may need to be changed. Similarly, it is more difficult for the exchange manufacturers to design different types of exchanges.

5.3.5 Switching Hierarchy

Exchanges in the PSTN are organized into a switching hierarchy. Not all exchanges have transmission links to every other exchange. Generally a four-layered structure is used, as shown in Figure 5.5.

Local

The customers are connected to the local exchanges. In some cases, the customers are connected via an RSU, but this is considered to be part of the local exchange, although it is physically located elsewhere. A local exchange will have transmission links to its *primary* center and perhaps to other nearby local exchanges.

Primary and Secondary Centers

Several local exchanges in the same area are connected to the *primary* exchange for that area. A number of primary exchanges are in turn connected to

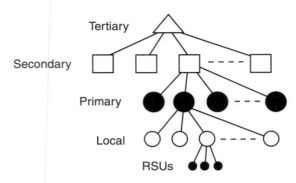

Figure 5.5 Switching hierarchy.

a *secondary* center. Finally, the secondary exchanges are connected to a *tertiary* center. Primary and secondary exchanges are normally large exchanges in the main towns and cities. Customers are also connected to the primary and secondary exchanges.

Tertiary Center

The tertiary center is normally the peak of the switching hierarchy. The tertiary center represents the main national and international exchange in the country. Because of its importance, normally no customers are connected directly to the tertiary center. In addition, usually two or more separate exchanges operate as tertiary centers, so that all national and international traffic can continue in the event of a failure in one of the exchanges.

In practice, a perfectly structured switching hierarchy is never actually implemented. As population density varies across a country, some areas may not have primary or even secondary centers. Then, the local exchanges are connected directly to the tertiary center. In urban areas with a long history of telephone use, the local exchanges may have so many interconnections with each other that a structured hierarchy becomes meaningless.

The advantage of having a structured switching hierarchy is that network management becomes easier, and it becomes feasible to introduce automatic network management procedures. Network management procedures are designed to ensure that the maximum number of calls can be successfully completed when the network is subject to exceptional traffic demands or serious faults.

5.3.6 Signaling

Signaling is the term used to describe the transfer of information required to control calls. For example, when setting up a call, the B number (i.e., the number of the called party) must be notified to the local exchange and any transit exchanges used for that call. *Interexchange signaling* refers to signaling between exchanges and *subscriber-line signaling* refers to signaling between the customer equipment and the exchange.

There are two categories of interexchange signaling: *channel-associated signaling* (CAS) and common-channel signaling.

Channel-Associated Signaling

The older interexchange signaling methods belong in this category. In CAS, the signaling information for a call is directly associated with the particular transmission channel carrying that call. Combinations of tones are used to send the call setup signals. For example, a particular combination of tones could represent the digit 6 in the B number. These tones are sent in the speech channel and can only be transmitted before the call is established. The most widely used tone system is called R2-MFC.

When digital transmission is used, some signals (such as *line seizure, answer,* and *release*) must be sent in a *signaling channel*. For example in R2-D (a signaling system widely used in Europe), time slot 16 of a 32-channel (2-Mbps) system is reserved for signaling information governing the 30 traffic-carrying channels on that system. Although carried in a separate channel from the speech channel to which it refers, each (4-bit) signal is uniquely associated with that speech channel.

The disadvantages of CAS are:

- A limited number of signals are possible, meaning that information essential for new enhanced services may not be carried.
- A signaling channel must be provided for every 30 speech channels, which is wasteful.
- Signaling and call traffic must travel the same route. This is not always desirable.

Common-Channel Signaling

In common-channel signaling, there is no fixed association between the channels used for signaling and the traffic channels to which the signals refer. By breaking this association, much more information can be transferred rapidly.

CCITT Signaling System Number 7 is the worldwide standard for common-channel signaling. The abbreviations *C7, SS#7,* and *CCS#7* are all used for the same system. C7 was designed according to the criteria of the *open systems interconnection* (OSI) reference model used for data communications. Although designed before OSI, the C7 signaling system becomes OSI-compatible when an optional part called the SCCP is added. In effect, C7 means that the processors controlling the exchanges are connected together using a fully featured data-communications protocol.

Instead of tones or 4-bit signals, the signals used in C7 are packets of data called *messages*. A wide variety of messages are specified, catering for a large number of services. In addition an *open* protocol is used, meaning that new groups of messages required for new applications can be added in the future.

Signaling messages are only sent when needed, meaning that a single 64-Kbps signaling channel can handle many more than 30 traffic channels. The protocols currently used allow a single signaling channel to handle over 4,000 traffic channels.

The signaling channels are grouped together to form a separate signaling network. This network can have alternative paths in event of failure or congestion on one particular signaling channel.

Signaling messages are not only associated with calls. Messages can be transferred relating to the state of the exchange equipment and the exchange itself. For example, an exchange that is experiencing a problem can inform the other exchanges in the network so that they can decide to route calls via an alternative route.

Finally, other types of equipment can be connected to the C7 signaling network. Examples of such devices are computers carrying roaming information (in mobile networks), voice-mail systems, and equipment produced by third-party manufacturers.

In summary, the main advantages of C7 signaling are:

- Capacity to carry information for additional services;
- Speed of call setup;
- Higher reliability;
- Ability to connect to other systems;
- Open design means greater flexibility to adopt new applications.

Subscriber-Line Signaling

Subscriber-line signaling refers to the signaling between the customer's equipment and the local telephone exchange. Until recently, subscriber-line signaling was very basic; essentially the only purpose was to transmit the digits of the B number.

Pulse Dialing. *Pulse* or *decadic* dialing was the subscriber-line signaling used in the first automatic exchanges. Modern exchanges still accept pulse dialing, although more modern systems are preferred. The digits are sent by means of a sequence of pulses, the number of pulses being equal to the value of the digit. The digit 0 is represented by 10 pulses. Pulse dialing has two main disadvantages:

- Dialing is a slow process—up to one second must be allowed for each digit.
- Pulses are not transferable past the local exchange, making access to additional services difficult or impossible. For example, remote access of a voice mailbox may not be possible.

Dual-Tone Multifrequency. DTMF, or more simply *tone* dialing, transfers the digits of the B number using a combination of tones. Each digit is represented by a unique combination of two tones. This system overcomes the two disadvantages of pulse dialing: dialing is faster and the tones can be used to access additional services and remote equipment such as voice mailboxes. In fact, DTMF is the standard used for most types of remote access and access to additional services. Telephones are available that use pulse dialing to set up a call (because the telephone is connected to an old exchange) but that change over to DTMF signaling once the call has been set up. Similarly, mobile telephones and ISDN telephones that use digital signaling will change over to DTMF when a call has been connected.

Recent Developments. Some new additional services require signaling information to be sent from the exchange to the customer's equipment. This is not possible using traditional pulse or DTMF systems.

One solution is the *custom local area signaling system* (CLASS) introduced by Bellcore in the United States. This enables, for example, the exchange to send the number of the calling party to the called party. In the United States this service is called *caller ID* or, more correctly, *calling line ID* (CLID). The number can be displayed on a CLASS telephone or on a stand-alone display unit. A similar system, *SIN 227,* is used in the United Kingdom.

Finally, modern exchanges allow digital connections. An example would be a first-order transmission system (2 or 1.5 Mbps) connecting to a PBX. In this case, a digital signaling system is used such as DASS2. DASS2 is a form of common-channel signaling and has the same types of features and advantages as C7. Similarly, in ISDN (covered in the next chapter), common-channel signaling is used, with DSS1 being the European standard.

5.4 The Local Access Network

The local access network is the part of the PSTN that connects the customer's equipment to the local exchange. Although the average length of an access is only about 4 km, the sheer number of accesses means that the local access network is the most expensive part of the PSTN. Typically, the local access network accounts for 50% of the investment in the network. Development in the local access network has been slow up to recent times; the traditional layout described in this chapter is largely unchanged from that used over 100 years ago. A hundred years ago, paper was used as the insulation on local cables, and lead was used as the protective outer sheath. Nowadays, polyethylene and PVC are used to insulate and protect the conductors in the cable, but little else has changed. In recent times, alternative layouts for the local access networks using radio, optic fiber, and CATV networks have been proposed and tested. However, these still account for only a very small proportion of total accesses.

The local access network is probably the most vulnerable part of the PSTN. Unless specifically requested, all of the lines connecting to a customer's premises will usually be carried on the same cables in the access network. Thus, if a failure occurs in one of these cables, that customer will be without any communication service. To avoid this, customers should request *diverse routing* in the local access network, so that their access lines are spread out on different cables. The ultimate in diverse routing is to have the access lines connected on separate cables and then terminated on two different local exchanges. A word of warning, though—simply because lines are terminated on different exchanges does not necessarily mean that diverse routing is implemented, as the lines could still be connected via the same cables in the access in the local access network (see Figure 5.6). Finally, expect to pay a lot for diverse routing. It is usually charged for on a case by case basis.

5.4.1 Traditional Layout

The traditional layout of the local access network is as shown in Figure 5.7.

The various elements of the traditional local access network are: local main cables, primary cross connection point, distribution cables, and distribution points.

Local Main Cables

These are cables containing between 200 and 4,000 pairs of wires. These cables are normally pressurized with dry air. This means that any damage to the cable can be automatically detected due to a sudden loss in pressure. In addition, the pressure of the air prevents moisture from entering the cable. Thus, damage to

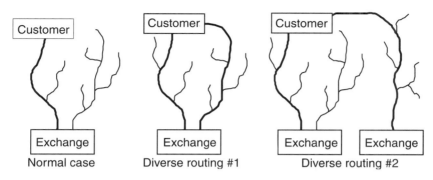

Figure 5.6 Diverse routing.

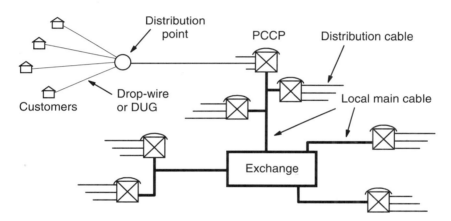

Figure 5.7 Traditional layout of local access network.

the cable can be repaired before the actual conductors are damaged. This means that local main cables are relatively reliable.

Primary Cross Connection Point

The *primary cross connection point* (PCCP) is a flexibility measure to allow changes in the local network without having to rearrange the local main cables. The local main cables are connected on one side of the PCCP and the *distribution* cables are connected to the other. By connecting *jumper* wires, any pair of wires in the local main cable can be connected to any pair in one of the distribution cables.

In areas of high telephone penetration and low growth, the PCCPs are normally omitted. In areas of low penetration and high growth, another layer of flexibility can be added in the form of *secondary cross connection points*.

The terms *flexibility point* (FP), *area distribution point*, and *cabinet* are all used to described PCCPs.

Early PCCPs were a source of a lot of problems in the local access network. Moisture in the form of condensation caused short circuits on the access lines. Inadequate termination methods caused problems, such as the tearing of the insulation on a cable pair, and after a few years the wires in the PCCP resembled spider webs! Modern PCCPs use *insulation displacement* (IDC) terminations, and these problems have virtually been eliminated. However, some of the earlier PCCPs are still in use.

Distribution Cables

These are the smaller cables connecting the customer's premises to the PCCP. Sizes range from 2 to 200 pairs. Distribution cables are not normally pressurized, so damage to the distribution cables are only noticed by manual inspection or, more normally, when the access lines on that cable go out of order and customers contact the network operator.

Distribution Points

The distribution point is the point from where the cables are connected into the customer's premises. For residential areas, a telephone pole or underground box can be used as the distribution point. For large business customers, cables are normally carried underground directly to the customer's premises, where they are terminated on a distribution frame.

A distribution point typically serves 10 access lines. Normally two pairs are cabled into a residential premises. These pairs can be contained in a *drop-wire* where the distribution point is a pole, or in a *direct underground feed* (DUG) in the case of an underground distribution point. The distribution point, drop-wire, and DUG feed are common sources of failure in the local access network.

Overhead Versus Underground

Cabled networks can be *overhead, underground,* or a mixture of both. Overhead cables are supported on telephone poles by means of a steel catenary wire in the cable. There is a limit on the weight that a pole can support—typically 4×175 pair cables is the maximum capacity of a standard pole. Underground cables are installed in a network of ducts (or pipes) and jointboxes. The limit on the number of cables depends on the number and size of the ducts. There is no weight limit, so underground cables can be any size. Cables containing 4,000 pairs are common.

There are advantages and disadvantages of both methods in terms of capital cost, maintenance, and capacity.

Capital Cost. Overhead cables are cheaper to install than underground cables. In particular, if there is no existing network of ducts, the costs of digging trenches, laying the ducts, and repairing a roadway can be extremely high. For this reason, overhead cables are used for long cables, such as in rural areas.

Maintenance. The fault incidence in overhead cables is significantly higher than in underground cables, so customers on overhead networks can expect a more regular occurrence of faults. For the network operator, the maintenance costs of overhead cables are higher. Overhead cables will also have a shorter life span due to greater exposure to the elements and accidental damage.

Overhead cables are more likely to suffer lightening damage. In addition, lightening can travel along the cables and damage the customer's equipment. Various methods of protecting against lightening damage can (and should) be employed.

One disadvantage of underground cables is that when a fault occurs in a section of cable midway in a duct, it can take longer to repair the fault. However, these types of faults are extremely rare, and most network operators will switch the line to a spare pair in the cable.

Capacity. As mentioned previously, underground cables can have a greater capacity than overhead cables. For this reason, underground is used for local main cables. Also, underground cables are used for the urban local access networks, although overhead is sometimes used for the distribution points and drop-wires.

5.4.2 Radio in the Access Network

Until recently, the use of radio in the access network was limited to a small number of high-capacity (2- or 1.5-Mbps) accesses. The widespread use of radio for normal capacity accesses is a recent development; in many cases, it is a result of competition in the local access network.

One of the complaints in this area is that existing monopoly network operators have deliberately used up radio frequencies for their mobile networks in a bid to prevent competitors from using radio.

The countries in which radio has been used to any major degree in the local network are Germany, Spain, the United Kingdom, and the United States.

Cost, speed of installation, and flexibility are reasons for implementing radio.

Cost. Radio costs are more or less independent of distance—unlike cable, where the length of the access is the most important factor relating to cost. For this reason, radio has been used for some time for very remote customers.

However, where a second operator is entering the local access network in competition to the existing operator, the cost of implementing a radio solution may be less than the cost of building an entire duct and underground network, even in urban areas.

Speed of Installation. Radio can be more rapidly deployed than cable. For this reason, radio has been used in many areas of the former East Germany to give rapid access to the network. It is also for this reason that second operators may decide to use radio in the local access network.

Flexibility. Radio networks are easier to reconfigure then cable networks. It is possible to remove radio equipment and redeploy it elsewhere. In the case of the former East Germany, where radio has already been used, the eventual solution may be to install a traditional cabled network, in which case the radio equipment can be rapidly redeployed in other areas.

Radio Standards

Multiple-Access Radio. *Multiple access radio* (MAR) is used to provide service to remote customers in rural areas. The reason for using MAR is cost, as the length of the accesses mean that radio is a cheaper solution than cable. This is the main example of the use of radio by established network operators.

Neighborhood Telepoints. Standards originally designed for cordless telephony or limited mobility systems can be used to implement a radio local access network using the *neighborhood telepoint* concept. The DECT and CT2 standards are the most often suggested. These are suitable for use in urban areas, as the range of the antennas is limited. For example, 100m is the normal range for DECT. The most prominent example is the DECT-based local access network in the center of Helsinki.

CT2 and DECT are covered in more detail in Section 9.5.4.

Personal Communication Networks

Personal communication networks (PCNs) are mobile networks with less mobility than standard mobile networks, such as GSM. The main technology is DCS 1800 (Section 9.5.1). The most prominent example is Mercury's One-2-One network in the United Kingdom, where the offer of free off-peak local calls is certainly in direct competition with the fixed network.

It is perhaps ironic that where the licensing of PCN operators has been delayed in an effort to protect the GSM mobile duopoly, future PCN operators may decide to target the market currently catered for by the fixed network.

It has been argued that a *de-engineering* of PCN technologies—such as DCS 1800, whereby the mobility element is more restricted or removed—could result in cost-effective technologies for implementing a radio local access network.

In Scandinavian countries, where mobile telephony has one of the highest levels of penetration into the market, it could be argued that the older (and less expensive) NMT 450 networks are in competition with the fixed network.

Fixed Radio Access. *Fixed radio access* (FRA) refers to a number of proprietary radio systems used to provide a radio local access network. These can be designed to operate at frequencies where there is spare capacity and, in particular, where they cannot be claimed by the mobile operators. The most prominent example is Ionica's network in the United Kingdom.

5.4.3 Optical Fiber in the Local Network

The main benefit of optic fiber is bandwidth. As more and more services, such as video on demand, multimedia services, and high-speed LAN access, are developed, the requirement for bandwidth is continually increasing. Only optic fiber can cater to the bandwidth required for these and future applications. Most analysts agree that the local networks of the future will be composed of optic fiber to the same degree that today's local networks are mainly composed of copper. However, the amount of investment required (remember that the local network accounts for about 50% of investment in telecommunications) means that ubiquitous fiber-optic local networks will not appear overnight.

There are two main scenarios in which optic fiber can be used in the local network, as shown in Figure 5.8.

Fiber to the Home

In certain cases, bandwidth—which is only possible using optic fiber—is required immediately. In this case, *fiber to the home* (FTTH) is deployed, whereby the complete access from the customer to the exchange is by means of optic

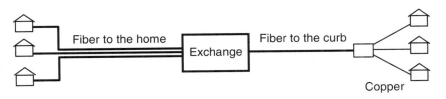

Figure 5.8 Fiber to the home and fiber to the curb.

fiber. The advantage is that the full bandwidth is available to the customer; the disadvantage is cost.

Fiber to the Curb

In many cases, the expense of cabling fiber all the way to the customer's premises is not justified, as only a portion of the bandwidth available over an optic fiber is required by each customer. *Fiber to the curb* (FTTC) involves cabling the fiber to a point outside the customer's premises. At this point, the fiber is terminated and the signals distributed by short copper cables into each of the customer's premises served by the scheme. The advantage is a minimization of cost.

As demand for bandwidth increases, the copper feeds can be replaced by optic fiber. However, the customers still share the one fiber back to the exchange. An example of such a configuration is the PONS system in the United Kingdom.

Eventually, when each customer requires the full bandwidth of optic fiber, a dedicated fiber can be installed from customer to exchange, resulting in an FTTH arrangement.

5.4.4 Cable Television

As youngsters growing up in Dublin, we referred to cable television as "the pipe." In those days, the pipe was used to bring us classic television programs such as "Top of the Pops," "Blue Peter," and "Coronation Street" from the United Kingdom. Nowadays, the same pipe can potentially be used to also distribute basic telephony services and even interactive multimedia applications such as home shopping.

Cable television (CATV) perhaps offers the greatest potential for competition on voice telephony in the local access network, in particular in the market for residential customers and small business users. At the moment, the only European examples of CATV networks providing voice services are in the United Kingdom. This is ironic, as the United Kingdom probably has one of the lowest levels of penetration for CATV in Europe, due in no small part to the excellence of its domestic terrestrial TV stations [2].

CATV Network Architecture

Many cable operators in the United Kingdom have used an overlay approach to providing telephone service in their networks. The *headend* of the television service is also used as the interconnection point for telephone services into the British Telecom and Mercury networks. CATV and telephony are carried together on optic fibers to lower level distribution nodes in the network. At the lower levels, the CATV signals are distributed by means of coaxial cables, and telephony is distributed by twisted-pair cables.

It is envisaged that digital transmission and compression of TV signals will eventually allow a single integrated connection directly to the customer's home.

References

[1] Minges, Michael, and Tim Kelly, *World Telecommunication Development Report 1996/97: Trade in Telecommunications Executive Summary,* International Telecommunications Union, http://www.itu.int/ti/publications/world/summary/wtdr96.htm, February 1997.

[2] "Laying the Lines for Local," *Network Europe,* March 1995.

Further Reading

Redmill, Felix, and A. R. Valdar, *SPC Digital Telephone Exchanges,* London: Peregrinus, 1990.

6

ISDN

6.1 Introduction

The *integrated services digital network* is essentially an enhancement to the ordinary telephone network—the PSTN. There are two main enhancements. First, the entire network is digital (in the PSTN, the telephone line and telephone are analog). Second, many new features have been added to the network.

The official birth of ISDN was in 1972, when ISDN was formally defined by the ITU. ISDN did not take off as a service until the 1990s. Much of the earlier problems were caused by several network operators interpreting the standards differently, with the result that equipment manufacturers were faced with considerable development costs to produce products for the different markets. As a result, equipment remained expensive and unattractive to most users.

A major step in overcoming the standards problem was the formulation of a pan-European standard called *Euro-ISDN-1,* which was inaugurated in December 1993. Since its implementation, reduced equipment costs and the high degree of compatibility between countries has meant a growing demand for ISDN connections in Europe.

A similar situation exists in the United States. Research company Bellcore produced a set of standards referred to as *National ISDN-1*. This has been less successful than the European example, as a number of prominent network operators refused to implement the common standards.

The ITU-T I-series recommendations are the ultimate source of much of the information contained in this chapter.

Opinions on the future of ISDN are divided. Some maintain that broadband technologies such as ATM are a much better option than ISDN. Others argue that ISDN is available *now,* albeit after a rather checkered history. They also point to technologies such as data compression, client/server, and inverse multiplexing, which make the bandwidths available over ISDN adequate for most applications.

6.2 Definition

The CCITT (now ITU-T) defined ISDN as "a network . . . evolving from a telephony Integrated Digital Network (IDN), that provides end-to-end digital connectivity to support a wide range of services, including voice and non-voice services, to which users have access by a limited set of standard multi-purpose user-network interfaces"[1].

6.2.1 Interpretation of Definition

The main points of the above definition of ISDN are explained next.

Evolved From the Telephony Network

The main feature of ISDN is that it is a development of the PSTN. The PSTN was originally an analog network, but over the years digital technology has been introduced, first in the transmission network and then in the switching system using 64-Kbps channels. ISDN is simply the extension of these 64-Kbps channels across the local access network to the customer.

As a result, a PSTN network operator needs only to upgrade its existing exchanges (central offices) to implement ISDN. Therefore, ISDN should be considerably less expensive to implement than an entirely new type of network. In particular, the local access network, which represents 50% of the investment in a network, is unchanged with the existing twisted-pair copper cables being used for the majority of ISDN lines.

The concept of evolution from the PSTN is also useful in trying to understand ISDN from the user's point of view. In fact, it is sometimes useful to consider ISDN as *PSTN version 2,* as it is basically the PSTN with faster bit rates. In particular, ISDN should have the same universal availability as the PSTN within a short time.

End-to-End Digital Connectivity

ISDN extends the 64-Kbps channels used in the digital PSTN as far as the customer premises. This enables end-to-end digital connections between ISDN users. The first advantage of this is that higher bit rates are possible—up to 128 Kbps on the same pair of wires as used for a normal PSTN line. When ISDN was first proposed, this was far greater than the bit rates possible on the PSTN using modems. Even now, the fastest modems using the PSTN can only offer up to 33.6 Kbps—about 25% of the capability of the same pair of wires connected to the ISDN.[1]

Two other advantages arise from the ISDN lines being end-to-end digital. Because the signaling system used to set up calls is now digital, calls can be established much faster. Once the call is connected, transmission can proceed immediately. If modems on the PSTN are used, it can take up to 30 seconds of *line probing* to determine the analog characteristics of the PSTN line before transmission can commence. In fact, when using ISDN, it is usual to be able to commence transmission of data within one second of dialing the called party's number. The other advantage of end-to-end digital connectivity is increased reliability—64 Kbps channels are guaranteed throughout the connection. There is never a case of having to fall back to lower speeds, as is often the case when using modems on the PSTN.

A Wide Range of Services, Including Voice and Nonvoice

The PSTN was designed to transmit voice signals in the range of 300 to 3,400 Hz. Any other type of signal had to be modified in some way—hence the need for modems when transmitting any kind of digital signal on the PSTN. ISDN was specifically designed to cater for any kind of service. Services such as voice telephony, video telephony, videotext, and telefax are defined and catered to by ISDN. In fact, the ISDN can even carry *unknown* services, yet to be conceived by equipment vendors.

Standard Multipurpose Interfaces

The advantages of standardization are obvious. In fact, the lack of standardization retarded the development of ISDN during the 1980s and early 1990s. At the moment, a high degree of standardization has been achieved on the interface from the network to the customer's equipment. However, in many areas the ways in which customers' equipment deal with the data do not meet the same standards. For example, a wide range of video telephones can success-

---

1. Modems operating at 56 Kbps in one direction are now available but will only work if one side of the connection is digital (e.g., PSTN to ISDN). The PSTN-to-PSTN speed limit is still 33.6 Kbps. High-speed modem specifications often claim throughputs of the order of 120 Kbps. This is possible using data compression techniques such as V.42bis (see Chapter 11); however, a 128-Kbps ISDN connection can deliver an effective throughput of up to 500 Kbps through similar compression techniques.

fully connect to the ISDN using the standard protocols. However, the way in which the images are encoded often differ, so it can be difficult to get equipment from different manufacturers to communicate successfully.

6.3 Access to the ISDN

In this chapter we refer to an ISDN line as an *ISDN access*. The term *access*, while less commonly used, is more precise because service can be provided over a radio link or using a portion of the capacity of a higher speed link to the central office.

The ITU-T definition of ISDN recommends two types of access: *basic-rate access* (BRA) and *primary-rate access* (PRA). In the United States, these are referred to as BRI and PRI—the "I" standing for *interface*. A BRA has two user channels, and a PRA has 23 or 30.

6.3.1 Channel Types

Bearer Channels

Normally referred to as simply *B-channels* (Figure 6.1), the bearer channels are the 64-Kbps channels used as the communication channels in ISDN. The different types of access differ in the number of B-channels offered. The B-channels can carry *unrestricted* 64-Kbps signals, in which case the ISDN network imposes no standards or limitations on the signals. In effect, the ISDN user can have full control over the use of the B-channels.

The Delta Channel

Normally referred to as simply the *D-channel* (Figure 6.1), the delta channel is the signaling channel used to control the communication. There is only one D-channel on each access, regardless of the type of access and number of B-channels. The separation of the control signals from the B-channels is what

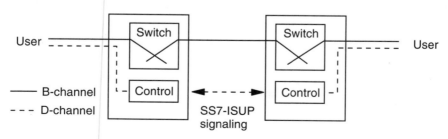

Figure 6.1 B- and D-channels in ISDN.

allows the implementation of unrestricted 64-Kbps signals. The signals on the D-channel, however, must adhere to the standard ISDN protocols. These signals take the form of *signaling messages,* which are used to indicate the called party number and type of services required.

For example, signaling messages are exchanged on the D-channel between the user equipment and the network when setting up a call. After a successful exchange of messages on the D-channel, the call is then connected on one of the B-channels. Other sequences of signaling messages are used to connect incoming calls, to disconnect calls, invoke supplementary services, and so on.

The bit rate of the D-channel depends on the type of access: BRAs have a 16-Kbps D-channel, whereas PRAs have a 64-Kbps D-channel.

The D-channel is normally used only for signaling, and cannot be considered as accessible to the user. However, some ISDN networks allow the D-channel to be used for access to X.25 packet switching (see Section 7.5), in which case the D-channel is accessible to the user.

6.3.2 Basic Rate Access

BRA is designed to be connected by the traditional twisted-pair copper wires used in the PSTN for normal telephone lines. To achieve the bit rate of 160 Kbps necessary to deliver a BRA, a special unit called a *network termination* (NT) is required at the customer's premises.

A BRA offers two 64-Kbps B-channels to the user. The two B-channels can be used for two independent calls or can be combined to create a 128-Kbps channel. This type of access is sometimes referred to as *2B+D.*

BRA is most suitable for domestic users and small business units. As an example, a banking company might decide to use BRAs for its smaller branches and the larger capacity PRAs in the head office and larger branches.

Installation and rental prices for BRA are normally about 1.5 to 3.5 times the prices for a PSTN line. As well as providing two separate communication channels, up to eight separate terminals can be connected to a single BRA. Each terminal can even have its own unique number if required. Thus, a single BRA may be a cost-effective alternative to a number of PSTN lines.

BRA is also suitable for connection to small PBXs. Five BRAs are the equivalent of 10 analog trunk lines. If the PBX is suitably equipped, ISDN capability can be extended to each of the extensions. The ISDN supplementary services can be used to extend PBX services, such as number identification, across the ISDN network to the remote end.

Finally, a number of BRAs can be used to provide a higher capacity data channel. The most common example of this is for videoconferencing, where 384 Kbps is the normal rate. Prior to the introduction of ISDN, this was only possible using leased lines. With ISDN, three BRAs can provide six B-channels with a combined capacity of 384 Kbps. When used for something such as LAN interconnection, a number of BRAs can provide a form of *bandwidth on demand*—for example, three BRAs can provide bandwidths of 0, 64, 128, 192, 256, 320, or 384 Kbps, which are only paid for as required.

Figure 6.2 shows the three main configurations using BRA.

6.3.3 Primary Rate Access

A PRA uses the same transmission format as used in PSTN transmission systems. Because there are two different transmission hierarchies, there are also two different formats for PRA:

- European format: 2 Mbps providing 30 B-channels and a D-channel;
- American format: 1.5 Mbps providing 23 B-channels and a D-channel.

PRAs are connected using optic fiber, radio, twisted-pair or dedicated PCM cables. When connecting a PRA to a customer, the network operator will normally use whichever transmission medium is most convenient. HDSL, which uses two or three "normal" PSTN lines to transmit 1.5 or 2 Mbps, has been in widespread use since 1996.

Typical installation and rental prices for a PRA are about 10 times the price of a BRA. Thus, PRA is only cost effective for larger business units or customers requiring high telecommunications capacity.

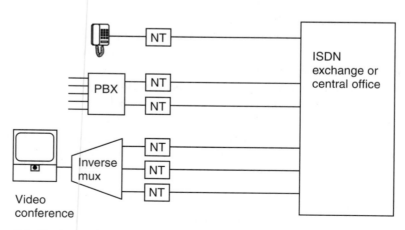

Figure 6.2 The three main configurations for basic rate ISDN access.

The majority of PRAs are connected to PBXs. If the PBX is suitably equipped, ISDN capability can be extended to each of the extension lines using the BRA standards.

6.4 Terms and Explanations

6.4.1 Reference Configuration

Figure 6.3 shows the reference configuration used for ISDN access.

The points R, S, T, and U are called reference points, that is, points at which the signals are standardized. The most important one is the S reference point, as this is the point at which most customers' equipment is connected. The line from the customer's premises to the exchange is called the U-interface, which is not defined by ITU-T.

The boxes in the diagram refer to *functions,* which are explained next.

6.4.2 Network Termination Type 1

Network termination type 1 (NT1) functions mainly concern line termination.

In the case of BRA, this involves the conversion of signals between the four-wire, 192-Kbps format used at the S-reference point and the two-wire, 80-kbaud (160-Kbps) format used at the U-reference point. The NT1 functions also provide multiplexing and timing for a BRA. In most cases, the NT1 functions for BRA are carried out by a small unit called a *network termination* (NT). In Europe, the NT is provided by the network operator and is considered as part of the network equipment. This is because the U-interface standards vary throughout Europe so a standardized NT is not possible. In America and Japan, the NT is considered to be customer's equipment. This has led to the development of equipment that combines the NT functions with other functions.

In the case of PRA, NT1 functions simply involve the normal line coding functions. This is normally carried out by a card on the PBX.

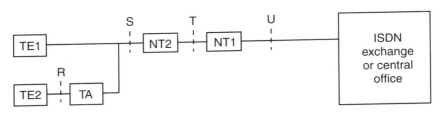

Figure 6.3 Reference configuration for ISDN (*From:* [2]).

6.4.3 Network Termination Type 2

Network termination type 2 (NT2) functions mainly concern switching and multiplexing. Thus, NT2 functions are carried out by PBXs or multiplexers.

NT2 functions are *not* required in the case of a simple BRA configuration, as switching is not required.[2]

Equipment providing NT2 functions is considered to be customers' equipment and can normally be installed by the customer or a third party.

6.4.4 Terminal Adapter

Terminal adapter (TA) functions convert the ISDN signals used at the S reference point to non-ISDN standards. TAs that convert from 3.1 kHz analog and older data interfaces (V.24[3]/V.28, X.21 and V.35) are the most common types of TAs used.

TA functions can be provided by stand-alone units, similar in appearance to modems. Many such units provide a number of possibilities—for example, one or two data ports and one or two *plain old telephony services* (POTS) ports. POTS ports can be used for ordinary telephones, fax machines, or even modems. Thus data terminals, personal computers, and PSTN equipment can all be connected via the TA to one BRA (Figure 6.4).

An important characteristic of a POTS port is the number of devices you can attach to it. This is measured in *ringer equivalencies* (REs). Typically the ports will support five REs. Most telephones have RE = 1, so you can connect five telephones in parallel (sharing the same port). Some terminal adapters, however, only have RE = 2.

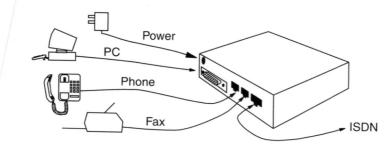

Figure 6.4 Multipurpose terminal adapter.

---

2. Because the NT2 is rarely found on a BRA, the S-bus is sometimes referred to as the S/T-bus.

3. V.24 is one of the most common types of data interface in use today. The serial port on PCs and modems are V.24

TA functions can also be built in to equipment. Many ISDN telephones include analog and V.24 TA functions. TAs can be provided as plug-in cards for personal computers. Many such cards also include analog and/or data ports.

In the United States, the possibility exists to purchase terminal adapters and other ISDN equipment with a built-in NT1. This is not the case in Europe, where the NT1 is owned by the phone company. While the built-in NT1 might save you money, it can in some instances prevent you from attaching other equipment, such as an ISDN phone to the ISDN line. This is because some devices with a built-in NT1 do not give you access to the S/T-bus (see Figures 6.5 and 6.6).

6.4.5 Terminal Equipment

Terminal equipment (TE) is classified as TE1 or TE2. TE1 refers to equipment that plugs directly into the S-interface and is fully ISDN compatible. Examples of TE1 are telephones, video telephones, and Group 4 fax machines specifically designed to be connected to the ISDN.

TE2 refers to equipment that was not specifically designed for connection to the ISDN. Examples of TE2 are analog telephones, modems, personal computers, and LAN equipment. TE2 must be connected via the appropriate TA.

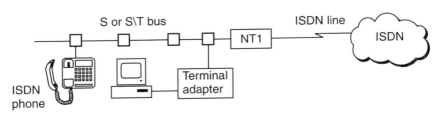

Figure 6.5 Typical European setup.

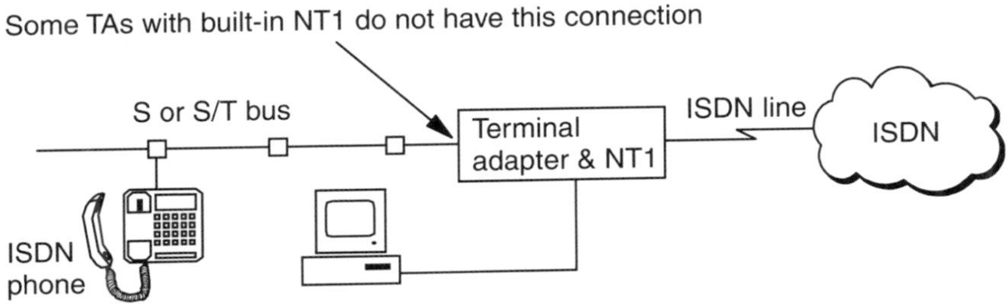

Figure 6.6 Typical U.S. setup with built-in NT1.

6.4.6 S-Bus

The ISDN standard allows for a number of TEs and/or TAs to be connected to the same S reference point. The cable used for these connections is referred to as the *S-bus* or *passive bus*.

The S-bus consists of a four-wire (most common) or eight-wire cable, to which RJ45 sockets are attached. The sockets are simple passive connections which are connected onto the cable in the same manner as traditional telephone sockets. The final socket on the S-bus, however, must be terminated with a 100-Ohm resistance between the transmit and receive pairs.

There are three configurations of S-bus recommended by the ITU-T (see Figure 6.7).

The *point-to-point* configuration allows the maximum distance from the TE to the NT, but only one TE can be connected to this type of S-bus.

The *short passive bus* configuration allows the maximum number of sockets to be connected (up to eight sockets) to the S-Bus, but imposes the shortest distance between the TEs and the NT.

The *extended passive bus* affords a compromise between the first two configurations, allowing a number of sockets to be connected at a medium distance between the TEs and the NT. However, in this configuration, the TEs must all be clustered within 50m of the end of the S-bus.

These configurations are designed to take into account two limitations on the signal transfer between the TEs and the NT—signal attenuation and timing delays.

An NT must be able to handle these configurations to meet the ITU-T recommendations for connection to an ISDN. However, an infinite number of variations are possible, as long as they meet the requirements imposed by attenuation and timing. NT manufacturers will usually publish a number of alternative configurations that have been tried and tested on their equipment.

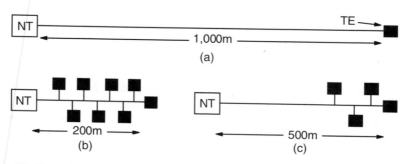

Figure 6.7 Three alternative S-bus configurations: (a) point to point, (b) short passive bus, and (c) extended passive bus (*From:* [3]).

6.4.7 Power for NTs and TAs

The normal situation all over the world is to supply power to the NT1 from a local source (i.e., you plug it in to the 110/230V supply). In the event of a power failure, an ISDN line may provide a limited amount of power. In North America, the ISDN line provides no power to the user. Outside North America, it can supply enough to power the NT1 and a single ISDN telephone. (Contrast this to a PSTN line, which can supply power sufficient for five or six telephones during a power failure.)

ISDN telephones usually draw their power from the S-bus. Terminal adapters usually require an external power source. Battery backup for NTs and TAs should be considered if operation during power failure is required.

6.5 ISDN Services

The types of services provided on ISDN are categorized as *bearer services, tele-services,* and *supplementary services.*

6.5.1 Bearer Services

Bearer services provide the ability to transmit certain types of signal.

The bearer service defines how the call is switched and transmitted through the network. The bearer service used for a particular call is determined by the initial *setup* message sent by the terminal on the D-channel to initiate the call. The most common bearer services are discussed next.

Speech, Circuit Switched.

In this case, the 64-Kbps B-channel contains normal PCM-encoded speech. The ISDN network can switch and transmit this call through the network in the same manner as a normal speech call. For example, the signal can be converted to analog and sent over analog transmission systems or the signal can be compressed using standard speech-compression algorithms. In most cases, the network operator will charge the same for this call as for normal telephone calls.

64-Kbps Unrestricted, Circuit Switched

In this case the information on the 64-Kbps B-channel contains information that must reach its destination unaltered. The network must ensure that the call is *not* switched or transmitted using systems that may be suitable for speech but would affect the information. For example, speech compression must not be used and the signal cannot be converted to analog. As these types of calls cannot use some of the existing network infrastructure, the network operator may need to install additional systems to cater to them. Thus, some operators will charge a premium of 25% to 50% above the normal telephony rates for these types of calls.

3.1-kHz Audio, Circuit Switched

In this case, the information on the 64-Kbps B-channel contains a PCM-encoded analog signal. This differs from the speech bearer service in that the signal may be voiceband data, such as from a modem or facsimile machine. The signal can be treated in almost the same manner as speech, but should not, for example, undergo speech compression.

56-Kbps Restricted, Circuit Switched

In this case, the information on the 64-Kbps B-channel contains digital information that has been rate adapted to 56 Kbps. Rate adaptation is done by ignoring the least significant bit in each byte (or octet). Thus, the call can be carried on T1 transmission systems that use *robbed-bit signaling*. These systems are common in the American and Japanese networks.

Packet-Mode Bearer Services

In this case, the call is a data call requiring connection through a packet-switched network. The packets can be carried either on the B-channel or the D-channel. The network operator will normally charge the same rates for ISDN packet-mode bearer services as for calls on the dedicated packet-switching network.

Reserved or Permanent Services

In this case the call is permanently set up through the network, providing the ISDN equivalent of a leased line on one of the B-channels. The network operator charges for this service in the same manner as for leased lines (i.e., by charging an annual rental).

Availability of Bearer Services

At the time of writing, the above bearer services are not all universally available. The speech, 3.1-kHz audio, and 64-Kbps bearer services are available in almost all networks, but packet-mode bearer services and reserved or permanent bearer services are only available on the most developed ISDN networks. In addition, other bearer services, such as *7-kHz audio,* are also defined but are not commonly available.

6.5.2 Teleservices

Teleservices are essentially the complete end-to-end communications capability.

The teleservice is specified in the initial *setup* message sent by the terminal on the D-channel to initiate the call. Each type of TE being programmed to send its particular type of teleservice in its setup messages.

The teleservice information is delivered by the network to the distant end where it can be used by the TEs at that end to decide whether the call can be accepted. The teleservice information is also used by the network to decide if tones (e.g., dial tone, call-back tone, busy tone) and announcements are to be used for the call.

Examples of teleservices are:

- Telephony (3.1 or 7 kHz);
- Teletex;
- Group 4 fax;
- Video telephony;
- Videotex;
- Teleaction.

With the massive changes in regulations since the ISDN definitions in 1972, the concept of teleservices has become increasingly irrelevant. An important teleservice for the 1990s is simply the "unknown" teleservice. Many ISDNs will handle the teleservice information *transparently,* which means that whatever the calling party sends as teleservice information will be sent on to the B-party unchanged.

6.5.3 Supplementary Services

Supplementary services are add-on facilities provided by the ISDN in addition to the call handling, bearer, and teleservices.

The range of supplementary services available varies depending on the network operator. The following are examples of the most common.

Number Identification Services

The identification of the calling party's number to the called party is almost universally available. The term *calling line identification presentation* (CLIP) is used to describe this service (*caller ID* is often used in the U.S.). Its use on international circuits is limited, however, due to varying national data protection laws, although its use for *call-back* applications may also have some bearing on this.

The term *calling line identification restriction* (CLIR) describes the service allowing the calling party to prevent his or her number from being presented to the called party. This can be used on a permanent or per-call basis.

Connected line identification presentation (COLP) describes the service allowing the presentation of the called party's number to the calling party. This is useful in the case where the called party is using a call-diversion service. The equivalent *connected line identification restriction* (COLR) is available to the called party if he or she wishes to prevent his or her number being presented.

Malicious call identification (MCID) services are also available to identify the calling and called numbers even where CLIR or COLR are used.

CLIP and CLIR are widely available. MCID, COLP, and COLR are less commonly implemented at the time of writing.

Direct Dial In

The DDI service (DID in the U.S.) enables a call to be connected to a PBX extension without the intervention or assistance of an operator. The extension number is included in the dialed telephone number. For example, the number 886–7124 could identify the extension number 7124 on the PBX with the number range 886–0000 to 886–9999. As far as the ISDN is concerned, DDI is achieved by sending the fourth and subsequent digits (i.e., the extension number) on the D-channel to the PBX. Obviously, the PBX must have the DDI facility to complete the call to the extension.

For outgoing calls, the handling of the DDI numbers depends on the ISDN operator. Some networks only allow the main PBX number (e.g., 886–0000) to be used for number identification services, such as caller ID. Some will allow both the main PBX number *and* the extension number to be used. Some will allow *either* the main PBX *or* the extension number to be used.

Charges for calls will be recorded against the main PBX number. Supplementary services are specified with respect to the main number. Typical DDI number ranges are based on blocks of 100 and 1,000 numbers. Usually, the DDI numbers must be a series of consecutive numbers.

DDI is widely available on both PRA and BRA. Many small businesses have switched to ISDN primarily to receive the DDI service, which is not available on PSTN lines.

Multiple Subscriber Numbers

The *multiple subscriber number* (MSN) service allows a range of numbers to be used on the one access, usually a BRA.

MSN numbers are used by the caller to call a particular terminal. This number is sent on the destination D-channel, and the particular terminal responds if it sees its own MSN. The other terminals ignore that incoming call. For example, an ISDN user might have a phone, PC, and fax machine connected to a BRA. Using the MSN service, the phone, PC, and fax can each be programmed with their own unique number.

For outgoing calls, the ISDN will allow any of the MSNs to be used for number identification services, such as caller ID.

Unlike DDI, MSNs are handled on an individual basis. For example, calls can be charged separately for each MSN and supplementary services can be applied differently to each MSN. Thus, the telephone could have call waiting, whereas the fax machine on the same line need not. The numbers used for MSN on a particular access do not need to be consecutive.

The nature of MSNs means that a large amount of data storage is required in the control processors in the ISDN exchange. Typically a maximum amount of 10 MSNs is allowed per access. If more numbers are required, customers are offered the more limited DDI service with its greater capacity.

MSN is relatively widely available. However, in some instances what network operators call an MSN service is, in fact, only the more limited DDI service.

Subaddressing

Subaddressing (SUB) allows the inclusion of more addressing information in addition to the number. This is handled by a subaddress field in the messages on the D-channel. In a manner similar to that described for MSN, subaddressing can be used to direct a call to a particular terminal in a group of terminals connected to the same access.

Subaddressing is more efficient for the network operator than MSN and is thus usually less expensive or even free. However, subaddressing is limited to calls originating within the ISDN, as PSTN and X.25 terminals cannot send subaddresses. Subaddressing is less widely available.

Call Diversion Services

These groups of services include call forwarding on busy, call forwarding on no reply, and call forwarding unconditional services. Call diversion—where the call is transferred to another number *after* it has been answered—is also possible. Call forwarding services are relatively widely available; call diversion is less widely available.

If true MSN is available, each number can be forwarded independently. Thus, it is possible to specify that voice calls are forwarded while data calls remain unforwarded. Some ISDN services allow individual call forwarding on the teleservice and/or bearer service (as opposed to the MSN). It will however be necessary to check this with your ISDN provider.

User-to-User Signaling

User-to-user signaling (UUS) services allow the transfer of packets of data on the D-channel. There are three versions:

- Type 1 (UUS1) allows the transfer of messages during the call setup phase (i.e., before or at the same time the number is dialed).
- Type 2 (UUS2) allows the transfer of messages during the alerting phase (i.e., while the call is ringing and has not yet been answered) and the call completion phase.
- Type 3 (UUS3) allows the transfer of messages during the conversation phase (i.e., after the called party has answered the call).

The contents of the user-to-user signals depend on the user's terminals. For example, user-to-user messages transferred by telephones are usually text messages, such as "call me back at 11:30." User-to-user messages transferred by file-transfer applications can include the file name and destination directory for the file being transferred. In some video-conferencing applications, user-to-user messages are used to send control signals to the remote camera (e.g., to "zoom in" or "pan left"). UUS1 is relatively widely available; UUS2 and UUS3 are less widely available.

Advice on Charge

This is a group of services providing information relating to the cost of the call, such as the number of call units and the cost per call units. Due to recent developments in the methods of charging, many operators have not implemented advice on charge services yet.

Closed User Groups

This involves the setting up of groups of users to and from which access is restricted. A particular user may, of course, be a member of a number of user groups (including the general public group). Closed user groups are not widely available at the moment of writing.

6.6 Applications

6.6.1 Introduction

The main feature of ISDN is the availability of switched 64-Kbps channels. This means that ISDN is suitable for many applications. We can define three categories of applications that benefit from the availability of a 64-Kbps channel [4]:

- The first category includes applications that benefit from increased speed but basically remain the same. This includes applications such as file transfer and email.

- The second category includes applications such as fax and Internet access, which benefit so much from the increased speed that they can be used in a new way. For example, at 64 Kbps, facsimiles can transmit pages at three to eight seconds a page, meaning that the transmission of complete documents is feasible. In addition, the quality of ISDN facsimile is similar to that achieved by a photocopier, making the transmission of artwork and "ready to use" documentation feasible.

- The third category includes applications that depend on 64-Kbps channels to operate effectively. This includes applications such as video telephony and high-quality speech. At present, the use of these applications at speeds below 64 Kbps does not appear to be practical.

In addition, a number of applications in the pre-ISDN environment could use 64-Kbps leased lines. This includes applications such as LAN interconnection and telesurveillance. Depending on the amount (and type) of traffic, it can be more economic to use ISDN for these applications. In certain cases, the improved economics of ISDN makes the adoption of these applications feasible to users who could not justify the costs of leased lines.

The number and range of applications using ISDN is changing all the time, so it is impossible to give an exhaustive list. The applications described in this chapter have been chosen to give an idea of the range of applications available.

6.6.2 Speech and Audio Applications

There are three possible ways in which ISDN can be used for speech and audio applications.

3.1-kHz "Normal" Quality Telephony

In this instance, the speech/audio signal is converted into a 64-Kbps digital signal using exactly the same PCM process as used in the PSTN (see Section 2.2.2). The only difference is that with ISDN, the PCM coding process takes place in the telephone rather than in the exchange. This offers a very slight improvement in quality, as analog noise in the local access network is eliminated. However, at least in the initial stages of ISDN development, it means that the telephone instrument costs about three times the cost of a similar PSTN telephone.

It would thus appear that the use of ISDN for 3.1-kHz telephony is not a viable option; however, this is not actually the case.

- First, ISDN PRA and BRA can be cost-effective options to PSTN lines for connection to PBXs. For example, a PRA may require a single termination card in the PBX, whereas 30 PSTN lines would require 30 termination cards. Additionally, the cost of a BRA may be cheaper than the cost of two analog PSTN lines.

- Second, the use of supplementary services such as CLIP, UUS, MSN, and DDI may justify the additional cost of the ISDN telephone or ISDN PBX.

- Third, ISDN access may be required for some other application. The integrated nature of ISDN means that the same ISDN line can also be used for telephony purposes. In this case the expense of the ISDN telephone can be offset against the savings achieved by retiring a PSTN line.

- Furthermore, the equipment used for the other applications may be complimentary with the telephony application. For example, many TAs have a built-in analog port, through which a standard PSTN telephone can be connected to the ISDN, and many ISDN telephones have a built-in data port, through which a PC can be connected.

As a result, 3.1-kHz telephones are the most common form of terminal equipment connected to ISDNs, although they are never the main reason the individual decided to use ISDN.

7-kHz High-Quality Speech

By using a different voice-coding technique to standard PCM, an audio bandwidth from 50 Hz up to 7 kHz can be encoded, with acceptable noise performance, using a 64-Kbps digital signal. This bandwidth matches the capability of the human voice to produce sounds.

At the moment, most users are content with 3.1-kHz telephony; however, some users require the extra quality available with 7-kHz bandwidth. The main examples are in the broadcasting industry, where 7-kHz speech is commonly used for news reports and sports commentaries. The advantage of ISDN over the use of leased lines is that charges relate only to the length of time that the line is used. It should be noted that the ability to use 7-kHz speech depends on the capabilities of the ISDN telephones at both ends of the connection.

Music Quality

Although human speech can only produce sounds from 50 Hz up to about 7 kHz, the human ear can typically hear sounds from 10 Hz up to 20 kHz. Thus, 7-kHz speech is not good enough for transmitting music.

To produce "hi-fi" quality sound requires not only a wider bandwidth, but also less noise in the signal. Thus, compared with standard PCM, more bits per sample are required at a higher sampling rate. A typical stereo compact disc, for example, produces a digital signal of 1.41 Mbps.

However, techniques have been developed for encoding this hi-fi bandwidth into lower bit-rate digital signals. For example, using the "Musicam" technique, CD-quality sound can be transmitted over 128 Kbps [4]. The main users for this application are within the broadcasting industry.

6.6.3 Remote Access and Internet Access

The advantages of accessing the Internet or a corporate network over 64 or 128 Kbps as opposed to normal modem speeds must be obvious. Besides simply being faster, the increased speed makes extra services feasible. For example, the visit of Pope John Paul II to the United States was available using "live" video to Internet users with ISDN access while Internet users with PSTN access had to make do with still pictures and a text commentary.

Add to this the advantage of rapid call setup times, and you can become easily convinced that ISDN is the only way in which you should access these networks. In fact, the demand for high-speed Internet access is one of the main driving forces behind ISDN development in the United States.

However, a few considerations should be taken into account first:

- The corporate network or Internet provider that you use must have an ISDN capability.

- The cost of TAs is still more expensive than modems. At the time of writing, the cheapest TA is still two to three times the cost of the cheapest 33.6-Kbps modem. However, prices are steadily reducing.

- In some countries, the cost of data calls on ISDN are more expensive than "normal" telephone calls. Thankfully, this is not always the case.

- In many countries ISDN is not as widely available as PSTN; problems are particularly acute in rural areas.

Terminal Adapters for Remote Access and Internet Access

Terminal adapters for remote access should ideally support the following features:

- The ability to combine both B-channels for a single data call, thus doubling throughput. There are many different schemes for combining B-channels, each designed for different situations. *Async inverse multiplexing* (AIMux) is used when the terminal adapter must emulate a modem so that the PC can use software designed for modems. Typically this is the case for access to bulletin boards or collaborative working. *Multilink PPP* (MLPPP) is used for Internet access and remote LAN access. *Bonding* is used with synchronous devices (e.g., routers, bridges, and video-conferencing equipment).

- The ability, while on a data call using both B-channels, to drop one B-channel to allow you to make or receive another call on that channel. The data connection remains active on the other B-channel, so you do not have to abort a large file transfer midway. (A proprietary extension to MLPPP [MLPPP Plus] supports this feature.)

- Battery backup (see Section 6.4.7).

- Front panel indicator lights to show what is going on (e.g., channels in use, line faults).

- Support for data compression. At the time of writing, many low-cost TAs did not support data compression. Even when compression is supported, the same compression technique may not be supported at the other end of the connection.

It should be noted that the data ports on many PCs (the serial ports) are not capable of the speeds that ISDN TAs are capable of—particularly adapters that combine B-channels and support compression. The solution to this problem is to use an internal adapter or one that connects to the PC's parallel port.

Terminal Adapters for the Central Site

The terminal adapters discussed so far connect to a basic rate ISDN line (BRA). If you need a large number of ISDN connections at a single site then primary-rate ISDN is usually more cost efficient than lots of basic-rate lines. A primary-rate ISDN line can be used to support 23 or 30 teleworkers dialing in at the same time.

Unlike basic-rate TAs, which often support POTS and data connections, primary-rate TAs are usually designed for data only. They are often designed with only two ports; one for connection to a LAN, and the other for connection to the ISDN line. They allow different callers to connect to different computers or servers on the LAN (Figure 6.8).

Such a TA is really a router (see Section 12.1). They are useful for allowing dialup access for teleworkers. They are also used by *Internet service providers* (ISPs) to allow ISDN dialup to the Internet. A useful but often expensive feature is built-in V.34 modems and a mechanism to determine whether a call is coming from a modem or an ISDN TA. With this feature, the one box can be used to support callers from both PSTN and ISDN lines (Figure 6.9).

6.6.4 LAN Interconnection

Prior to ISDN, the most common communication service used for interconnecting LANs was leased lines. However, ISDN is in many instances a more economic alternative.

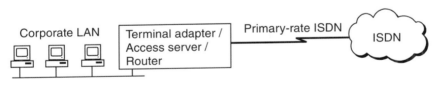

Figure 6.8 Connecting a LAN to a primary-rate ISDN line.

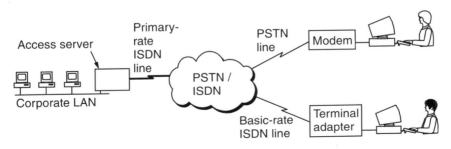

Figure 6.9 Teleworkers dialing into the company LAN.

The nature of traffic between LANs is usually bursty. Thus, it may be more economic to pay ISDN *call* charges rather than the *rental* charges for leased lines. The capacity of leased lines and ISDN is the same, each offering 64-Kbps channels up to 2 Mbps.

To take advantage of the economics of ISDN, calls should be set up only when data needs to be transferred. The call setup times for ISDN are sufficiently short to allow this without aggravating users. A potential problem exists, however, if management messages are not restricted. For example, servers regularly issue broadcast messages and if this activity was left unchecked, it could generate an ISDN call every 30 seconds. The way around this problem is to ensure that *all* of these messages are *spoofed.* An inefficient or absent spoofing method can result in increases in ISDN call charges of the order of several thousand percent! Spoofing is described in Section 12.2.1 and in [5]. The cost of the extra time required to configure and reconfigure ISDN LAN interconnect equipment should not be overlooked.

6.6.5 Video telephony

In the past, video-telephony applications have been implemented at very low quality using PSTN modems or at very high cost using leased lines. ISDN now offers a high-quality alternative that is more economic than leased lines. By combining B-channels, any desired video quality can be achieved. Video telephony is further described in Section 11.7.

6.6.6 File Transfer

The main advantage in using ISDN for file transfer is the increased speed over using the PSTN. This is achieved most obviously in the availability of a bit rate of 64 Kbps. Less obviously, considerable time is saved in the call setup. *Handshaking* over ISDN takes less than four seconds, whereas handshaking between modems on the PSTN can take up to 30 seconds or more. The second advantage is the reliability of the ISDN channels, particularly on international calls.

Although ISDN can be more expensive per second for calls, the increased speed means that the calls are connected for a much shorter time and therefore can cost considerably less than PSTN calls.

Often, the file transfer application also includes voice. This is referred to as a voice and file transfer application. This is made possible by allowing one B-channel on a BRA to be used for file transfer at the same time as the other B-channel is used for a voice call. Thus, the users can talk to each other as the file transfer takes place.

6.6.7 Group 4 Facsimile

Facsimile is the second-most-used service on the PSTN after telephony. The standard used is called Group 3. Group 4 facsimile, strictly speaking, refers to an improved way of encoding the information to be transferred and can be used on any kind of transmission medium. However, most Group 4 facsimile machines are designed for connection to a 64-Kbps channel, such as the B-channel in ISDN [4].

The advantage of Group 4 facsimile is the capability of sending higher resolution, and therefore better quality, images. When used on ISDN, a further advantage is the speed of the 64-Kbps B-channel in comparison with 9,600 bps, which is the most common speed used by Group 3 facsimiles on the PSTN. As discussed earlier, the increase in speed is of such an order that the way in which we use fax machines may change. Group 3 fax machines are generally used to transmit short pieces of information; with Group 4 fax machines it becomes feasible to transmit complete documents.

As with Group 3 fax, Group 4 fax can be implemented by means of a dedicated fax machine or by a plug-in card for a personal computer. Dedicated Group 4 fax machines are normally plain paper machines and can usually be identified by larger paper trays than Group 3 machines. A number of options are generally available, allowing the user to choose between higher resolutions and faster speeds. Unfortunately, these extra features mean that at the moment, Group 4 fax machines are about 10 times more expensive than Group 3 machines.

For this reason, there are not too many around. There is no point having one if the people you wish to communicate with do not. In ISDN jargon, the critical mass has not been reached. It is also possible (but not common) to have Group 4 fax capability built into terminal adapters.

It can be argued that a Group 4 fax machine can pay for itself if you have a high volume of international faxes to send. With Group 4 fax, the calls will be 5 to 10 times shorter and therefore a lot less expensive. However, many organizations are able to cut their fax bills in a much more dramatic way through the use of email or file transfer. It is quite possible that Group 4 fax will never reach critical mass.

6.6.8 Database Indexing

This is an application which makes use of the CLIP supplementary service. The same service is sometimes called *automatic number identification* (ANI) or *calling line ID* (CLID).

Essentially, database indexing involves cooperation between a company's PBX and its computer system (see CTI in Section 3.5.7). When someone dials the company, the caller's number is used to automatically search the company's database. For example, a customer-support line could use database indexing to present the customer's details to an operator at the same time that the call is delivered. A further development is for the computer system to analyze the details in the database and then instruct the PBX to connect the call to a particular person. For example, for a technical-support service, the call could be connected to the department dealing with the particular item that the customer had recently purchased.

The advantages of database indexing derive from the customers' calls being dealt with more quickly than before:

- Reduced costs for *freephone* (800) services, where the receiver pays for the calls;
- Reduced operator costs, as each operator can handle more calls per hour;
- Increased customer satisfaction, as queries are handled more efficiently.

Database indexing can work on the PSTN in countries that have calling line identification services on PSTN lines. This is true in most of the United States.

6.7 Interworking

6.7.1 ISDN–PSTN Interworking

Because ISDN is an evolutionary development of the PSTN, interworking between the two networks is simple. In fact, a fully developed ISDN and the PSTN are the same network, with some customers having access to all bearer services while others have only analog access to the 3.1-kHz audio bearer service.

Of course, as PSTN customers only have access to 3.1-kHz audio, this is the only bearer service that will interwork from PSTN customers to ISDN customers. In the other direction, 3.1-kHz audio and speech bearer services will interwork. Thus, there is no problem in interworking telephony service between ISDN and PSTN customers.

Data services (including fax and video), which in ISDN would normally use the 64-Kbps or 56-Kbps bearer services, are another story. These services can only interwork using the 3.1-kHz bearer service, with the data being encoded into a voiceband signal using modem standards. Fortunately, this is achievable by either of two methods:

1. In the simplest case, the ISDN customer installs a *voice-to-S* TA. This effectively provides a PSTN standard port on the customer's ISDN access, to which a modem or PSTN fax can be connected. The ISDN customer can then make and receive PSTN calls. Some of the latest ISDN cards for PCs include a *digital signal processing* (DSP) chip, which can emulate a modem and voice-to-S TA. This enables the ISDN customer to make and receive both ISDN calls (using 64 or 56 Kbps) and PSTN calls (using modem standards over 3.1-kHz audio) using the same card.

2. Some networks provide a bank of modems and adapters in the ISDN exchanges. By dialing a prefix before the telephone number, an ISDN customer can be connected via one of these modems to a PSTN customer. Similarly, a PSTN customer can access ISDN customers. This is not universally available.

Obviously, as well as being able to achieve a connection, the standard of the equipment at the PSTN and ISDN ends of the call must be compatible. In general, most ISDN equipment can drop back to the equivalent PSTN standards. Group 4 fax machines can usually drop back to the Group 3 standard, and 7-kHz audio equipment can drop back to 3.1-kHz audio. Video-telephony standards are still evolving, but the intention is that video-telephony equipment on the ISDN will be able to drop back to the video-telephony standards used on the PSTN.

6.7.2 ISDN–X.25 Interworking

There are four main ways in which an ISDN customer can access an X.25 network.

3.1-kHz Audio Bearer Service Interworking

The previous section in this chapter, ISDN-PSTN interworking, explained that ISDN customers could make data calls to the PSTN by using a modem and voice-to-S TA or an ISDN card with a DSP chip. In the same way, ISDN customers can access an X.25 network as if they were accessing it from the PSTN.

64-Kbps (or 56-Kbps) Bearer Service Interworking

If provided by the X.25 network operator, an ISDN customer can access the X.25 network using the 64-Kbps or 56-Kbps bearer services. This is sometimes referred to as *case A* access. Compared with *case B* access, this has the disadvantage of using one of the B-channels on the ISDN access for the duration of the connection to the X.25 network.

Packet-Mode Bearer Services Interworking

On some networks, the D-channel of an ISDN access can be used to allow a packet-switched access to an X.25 network. This is sometimes referred to as case B access or *X.25 on D-channel*. When using this bearer service, the customer's equipment sends the packets on the D-channel. At the ISDN exchange, these packets are frame relayed directly to the X.25 network. This method has the great advantage that it does not use any of the B-channels, which remain free to be used for other types of calls. As it is more efficient it should, presumably, be less expensive. The main (technological) disadvantage is that for BRA customers, a maximum throughput of about 9,600 bps is achievable on the 16-Kbps D-channel. However, this data speed is perfectly adequate for most packet-switching applications. However, of more importance are the main operational disadvantages:

- Where provided, case B access is normally only provided to the network operator's X.25 network. Case B access to a competitor's packet-switched network is rarely available.

- Case B access is not widely available. Many network operators have been slow to implement it (as of 1997).

Greater availability of case B access on ISDN could extend the life of X.25 packet-switched networks. For example, the cost effectiveness of packet switching, coupled with the rapid call setup times of ISDN, is ideal for applications such as credit card verification.

Semipermanent Bearer Service Interworking

ISDN customers can be semipermanently connected to an X.25 network. This is an alternative to a leased-line connection to the X.25 network. In effect, one of the B-channels of the ISDN access becomes a 64-Kbps leased line connecting to the X.25 network. The other B-channel(s) can still be used as normal for circuit-switched calls. However, the same operational disadvantages as mentioned for packet-mode bearer services apply. Semipermanent access is not universally available, and where it is available it is usually only to the network operator's X.25 network.

6.7.3 International ISDN Interworking

In the early days of ISDN, interworking between ISDNs of different network operators and, in particular, across national boundaries was a major problem. This gave rise to the phrase "islands of ISDN," because ISDN users could only connect to other users on the same network. Since then, however, some "bridges" have been built between these "islands."

Signaling

The main factor governing interworking between ISDNs is the signaling system used between the exchanges. Essentially, the interexchange signaling must be able to convey information relating to the bearer service, supplementary services, and teleservices being used for each call. Older signaling systems are not sufficiently flexible to allow this information to be transferred, because they were designed with PSTN services in mind. For example, if an ISDN call requests a 64-Kbps bearer service, then the call must be carried on digital circuits from end to end. Older tone signaling systems cannot pass on the request for a digital circuit, so even though sufficient digital circuits may be available in the destination network, the call is treated as a normal PSTN call and may be routed over analog circuits.

The signaling systems used for ISDN interconnections are variants of a digital signaling system called *signaling system number 7* (SS7). As this was developed from CCITT recommendations, it is also sometimes called C7. The first version of SS7 was designed for the PSTN, and is called *telephone user part* (TUP). Two later version of SS7 are used for ISDN.

TUP-j

A minor modification of TUP, called TUP-j, is suitable for ISDN interconnections. This provides the ability to set a bit (the *j bit*) in one of the messages to indicate whether a 64-Kbps circuit is required end to end.

TUP-j, however, is very limited in that it cannot transfer information relating to the supplementary services, teleservices, or other bearer services besides 3.1-kHz audio and 64-Kbps unrestricted. It is generally only used as an interim measure to provide a basic ISDN international service until *ISUP* signaling can be implemented.

ISUP

ISDN user part (ISUP) is a variant of SS7 designed particularly for ISDN. It is capable of transferring all of the information required to control ISDN calls.

Where networks are connected using ISUP, there should be no restrictions due to signaling.

Network Compatibility

The other main factor governing the effectiveness of ISDN interconnections is the compatibility of services offered by the different ISDNs. Obviously, if the destination ISDN does not support a particular service supported by the originating ISDN, then this service cannot interwork.

One of the problems discovered by users of international ISDN is that the setup of equipment and the way it is used can differ from country to country. For example, one multinational company reported that it took far too long, about 10 seconds, to set up an ISDN call in Spain, whereas there were no problems in other countries. When they investigated, they discovered that phone numbers in Spain should be ended with a special "end of number" character, which solved the problem. Another company reported problems when an ISDN backup for a leased line failed to work in the Netherlands, whereas the equipment had previously worked on the German network. Investigations revealed that the Netherlands network turns off the power to ISDN circuits that are not in use and must be activated by a pulse before a call can be connected [6].

In summary, these experiences show that when using international ISDN, it is best to source equipment and expertise in the country in which it will be used.

Terminal Compatibility

Of course, even where the networks are compatible and use the latest version of ISUP signaling, a call can only be successful where the terminal equipment at either end is compatible. This can be a problem nationally, but becomes even more common with international use. Telephony and fax standards are universally compatible. The main problems can occur with video-telephony and file-transfer applications where a number of different standards exist.

6.8 ISDN Development and Deployment

6.8.1 Development of an ISDN Network

To evolve from a PSTN to an ISDN, the following modifications to the network are required:

- Install software in the exchanges to handle ISDN services;
- Install ISUP (or at least TUP-j) signaling between the exchanges;
- Install interfaces in the subscriber switching stages (see Section 5.3.2) of the exchanges for BRA and PRA.

In addition to the above, the local access network may need to be modified, although the standards are designed to use the existing local access network.

Because of the size of the PSTN, the implementation of an ISDN is often by means of an *overlay* network. As demand for ISDN increases, the PSTN exchanges can be modified to provide an *integrated* ISDN/PSTN (Figure 6.10).

Overlay Network

It is possible to locate the subscriber stages remotely from the main exchange. These remote switching stages are called *remote subscriber units* (RSUs) and can be located at any distance from the main exchange.

In an overlay network structure, a small number of ISDN exchanges are installed in the larger cities. ISDN service is provided in the smaller cities and

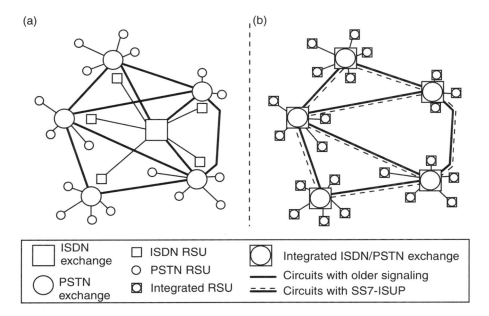

	ISDN exchange		ISDN RSU		Integrated ISDN/PSTN exchange
	PSTN exchange		PSTN RSU		Circuits with older signaling
			Integrated RSU		Circuits with SS7-ISUP

Figure 6.10   (a) Overlay and (b) integrated ISDN networks compared.

large towns by installing RSUs, which are connected to one of these ISDN exchanges. Thus, implementation of an ISDN network is faster because the number of ISDN exchanges are kept to a minimum. In particular, the number of routes requiring ISUP signaling is consequently smaller.

Areas with a high demand for ISDN are served by one of the ISDN exchanges or by an RSU connected to it. Alternatively, service can be provided to areas with a low demand for ISDN by using the digital transmission network. Primary rate can be directly connected on the transmission network, as it uses the same bit rate and frame structure. Basic rate can be multiplexed, using a special multiplexer called a *B-Mux,* onto the frame structure used by the transmission network. In Europe, for example, 12 basic rates can be multiplexed onto a standard 2-Mbps transmission system.

The advantages of an overlay network are:

- It is the fastest way of deploying wide availability of ISDN.
- ISDN exchanges means less possibility for variation in setup of equipment.

The disadvantages of an overlay network are:

- *Vulnerability:* If one of the ISDN exchanges fails, a large area will be without ISDN service. In general, because ISDN is a relatively new service, the reliability of ISDN exchanges is less than for PSTN exchanges. For areas served by RSUs and B-Muxes, there is a greater than normal dependency on the transmission network, as it is not possible to implement alternative routing.
- *Numbering:* Because ISDN service is provided from a different exchange than the PSTN exchange, the number range for ISDN must be separate from the PSTN number range in a given area. The main disadvantage is that customers who upgrade from PSTN service to ISDN must be given new numbers.

Integrated ISDN/PSTN

In this case, all of the digital exchanges in the PSTN are upgraded to provide ISDN services, and sufficient routes are upgraded to use ISUP signaling. The ISDN and PSTN are now the same network, simply with three different types of access: analog, basic-rate ISDN, and primary-rate ISDN.

An overlay network can evolve into an integrated network, simply by *reparenting* the ISDN RSUs to the local exchanges after they have been upgraded. Obviously, the process of reparenting can take some time, depending on the size of the network.

Once the local exchanges have been upgraded, the existing PSTN RSUs can also be upgraded to provide ISDN interfaces. Thus, the availability of ISDN can penetrate further into the rural and suburban areas. The cost and ease with which these PSTN RSUs can be upgraded to ISDN working depends on the manufacturer. In the simplest case, it merely involves the installation of a single card in the RSU—therefore, once the ISDN and PSTN are integrated, ISDN becomes available to all customers served by these types of RSU, even in the most remote areas. In other cases, it involves the installation of a new *shelf* of equipment. Thus, the network operator may be reluctant to install a shelf in an RSU unless there is sufficient demand in the area to justify the cost. In this latter case, availability will be restricted in rural areas served by these types of RSUs, unless demand for ISDN increases.

In most networks, a portion of the network will remain *overlay* for some time, as the local exchange cannot be upgraded to ISDN. For example, electro-mechanical exchanges cannot provide ISDN.

6.8.2 Functionality

As ISDN develops in a country, the amount of services provided by the ISDN increases.

International Connectivity

For ISDN to work fully between exchanges, ISUP signaling must be used. TUP-j signaling can provide only a limited ISDN interconnection. Within the operator's own ISDN, it is relatively easy to arrange for ISUP signaling between the exchanges; however, the task becomes more complicated when dealing with international connections.

Thus, the amount of international connections will increase gradually over time, so those considering the use of ISDN in different countries must investigate this. There are a number of factors to bear in mind:

- Check whether the signaling is ISUP or TUP-j, as this will dictate the types of services that can be used over the international links. Even if ISUP is used, check the standard, as some older versions of ISUP are more limited.

- Make no assumptions! In particular do not assume that because there is an international connection from X to Y that there must be an international connection from Y to X.

- Investigate surcharges for international *data* calls. These can range from zero to over 200%. Some operators apply heavy surcharges for international cross connects. Thus, even though many other operators may have a policy of not applying data surcharges in their national networks, they may be forced to surcharge international calls where they use an international operator that does apply them [6].

- Check CLIP interworking. There are often restrictions on transmitting CLIP information on international circuits, due partially to data-protection laws. The arrangements do not usually apply in both directions. Whether CLIP is sent can depend on the data-protection laws in the destination country.

Bearer Services

In the initial phase of development, normally only the 3.1-kHz audio, 64-Kbps unrestricted, and 56-Kbps restricted bearer services are offered.

In later phases, additional more specialized services, such as X.25 on the D-channel and $N \times 64$-Kbps unrestricted bearer services, are offered.

Supplementary Services

Initially, a limited number of supplementary services are offered. For example, phase 1 of Euro-ISDN specified that terminal portability, CLIP, CLIR, DDI, and MSN needed to be available.

As exchange software develops and standards are agreed, other supplementary services are made available. Subaddressing and user-to-user signaling are often available quite early on. Basic call-forwarding services are usually next to become available. More advanced services such as COLP/COLR, malicious call tracing, advice on charge, closed user groups and the more advanced call diversion services become available later.

References

[1]   CCITT, Recommendation I.110, "Preamble and General Structure of the I-Series Recommendations for the intergrated Services Digital Network (ISDN)," (Malaga-Torremolinos, 1984; amended at Melbourne, 1988), ITU, 1988.

[2]   ITU-T, "I.411 ISDN User Network Interfaces—Reference Configurations," ITU, 1993.

[3]   ITU-T, "I.430 Basic User Network Interface: Layer 1 Specification," ITU, 1995.

[4]   Griffiths, John M., *ISDN Explained Worldwide Network and Applications Technology,* 2nd ed., Chichester, England: John Wiley & Sons, 1992.

[5]   Heywood, Peter, "Ouch! 10 tips to keep ISDN from hurting your budget and your career," *Data Communications International,* November 1996.

[6]   Heywood, Peter, "ISDN: A long time coming, but still no quick fix," *Data Communications International,* November 1995.

7

Public Data Networks

7.1 Introduction

In Chapter 4, we saw how LANs can be used to carry data within a building. We will now look at networks that carry data over longer distances including networks that can span the globe. These networks are often referred to as WANs. The Internet is the biggest data network in the world and as such has been given a chapter on its own; however, because most readers will already be familiar with the Internet, it is used in some of the examples in this chapter.

Data networks can be either public or private. Although this chapter deals mainly with public data networks, private data networks can be built using the same or similar technologies combined with lines leased from service providers such as telephone companies or satellite consortia.

7.1.1 Uses of Data Networks

Public data networks can be used in a variety of ways. First, they can be used to interconnect the computing facilities (LANs, mainframes, terminals) in the different locations of a particular company or corporation (see Figure 7.1).

Second, they can be used by a company that wishes to allow public access to its computing facilities—for example, to allow its customers to log on to them and browse databases, download files, or use its email services. This type of company is referred to as an information provider or a service provider.

Finally, data networks can be used by individuals or companies who want to gain access to the facilities provided by information providers or service providers. Many of these individuals will gain access to the data network by

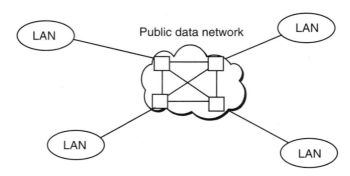

Figure 7.1 Interconnection of LANs via a public data network.

means of a local call over the PSTN (the ordinary telephone network), using a modem (Figure 7.2).

Listed next are a number of services that can be provided over a public data network. It should be noted that many of these services can be and are also provided by means of dialup connections over the ordinary telephone network.

- *Internet access:* The Internet is itself a public data network; however, it is often the case that a company's connection to an *Internet service provider* (ISP) is made via another data network;

- *Email:* Although email is invariably offered along with Internet access many service providers offer other forms of email (such as X.400 mail). These mail systems usually have access to and from the Internet because Internet mail has become the de facto email standard;

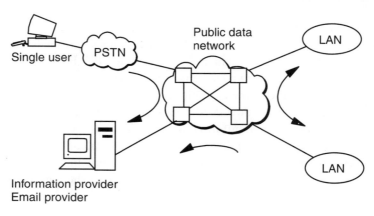

Figure 7.2 Access to a service provider.

- *Information services:* online encyclopedias, financial information, mailing lists, directories, and software bulletin boards;
- *Online banking;*
- *Online shopping* (catalogs and order forms).

Many modern data networks can be used to carry digital voice and digital video in addition to what one normally associates with the word *data.*

7.1.2 Data Over the PSTN

The PSTN can be used quite effectively to carry low-volume data. Its use has been encouraged by the many dialup services available over the PSTN. Sending data over the PSTN is achieved using modems, as described in Section 11.2. The main drawbacks of using the PSTN are:

- It is expensive for long-distance calls.
- Charges are based solely on call duration.
- Call setup and modem handshaking take a long time (typically 20 to 30 seconds).
- The maximum speed is limited to around 30 Kbps (56 Kbps in some cases).
- Computer systems connected to the PSTN are quite exposed to hacking, as the PSTN is so widely available and has little built-in security.

For these reasons the PSTN is not an ideal solution for large-business users who wish to interconnect their internal computer systems. The PSTN is often quite appropriate for gaining access to a service provider, particulary ISPs. It is also used quite extensively for remote access to a company's computing facilities. In this latter case, security measures must be taken to reduce the exposure to unauthorized access.

7.2 Leased Lines

Although leased lines cannot be classified as a data network in themselves, they are nonetheless a very important element in building private and public data networks.

A leased line (private line in the U.S.) is a permanent private communications link between two locations. It is normally available 24 hours a day, and the capacity on the link is exclusive to the owner. It should not, however, be visualized as a dedicated cable between the two locations. Rather, it is a dedicated portion of the bandwidth in the various cables along the way. In some rare instances, dedicated cables can be leased, such as when an organization leases optic-fiber cable links to build an FDDI backbone between adjacent buildings.

Leased lines can be broadly categorized as digital or analog. While either type can be used to carry voice or data, analog lines are more suited to voice while digital are more suited to data or voice-data integration. Analog leased lines carry data through the use of modems similar to those used on PSTN lines. The maximum speed is limited to 33.6 Kbps, but this speed is not guaranteed.

7.2.1 Digital Leased Line Speeds

The main differentiating factor between digital leased lines is the speed of operation. The lowest speed is normally 64 Kbps in Europe and 56 Kbps in North America. (Transatlantic lines often work at 56 Kbps.) In North America, the next speed up is often a full T1 circuit of 1.5-Mbps capacity. Many carriers can, however, offer intermediate rates using *fractional T1*. In many European countries, intermediate rates such as 128 Kbps and 192 Kbps are available as *N × 64 Kbps* circuits in addition to the E1 rate of 2 Mbps. The next step above T1 and E1 is normally T3 (45 Mbps) and E3 (34 Mbps)—quite a jump in capacity. Needless to say, very few organizations can afford T3 or E3 capacity.

7.2.2 Leased Line Reliability

A very important characteristic of a leased line is its reliability, normally measured in terms of percentage availability and maximum time to repair. The providers' backbone transmission networks are normally protected with redundant transmission capacity, which gets switched in automatically in the case of a fault. The local loop, however, is not normally protected by a redundant circuit, and the customer will have to pay a premium to get this protection (see Section 5.4).

If reliability is particularly important, the option of using a satellite link as opposed to a terrestrial link might be considered. They are often more reliable, particularly for international links. Another means of getting high reliability in some countries is to obtain a leased line based on a self-healing SDH ring.

7.3 Data Networks in General

7.3.1 Variety of Data Network Types

Although voice networks, such as the telephone network, can be and are used to carry data, special data networks are often faster or more cost effective for this purpose.

In the future it should be possible to carry voice, data, and video efficiently using a single network technology—namely ATM—but for the present we not only have separate voice and data networks, we have many different types of data network:

- Circuit-switched data networks;
- X.25 (packet switching);
- IP networks (e.g., the Internet);
- ISDN;
- Frame relay;
- SMDS;
- ATM.

Each of these network types has its own particular characteristics. The earlier networks were not particularly good at integrating different types of information (voice, video, and different types of data) so we have ended up with different networks for different applications. As technology develops, it becomes easier to integrate different traffic types on the one network.

7.3.2 Nature of Data Traffic

How is data so different from voice?

Bursty

Data transmission often involves periods of activity followed by periods of inactivity (e.g., somebody looks up some information and then reads the screen before requesting more information). This type of data is referred to as *bursty* because it consists of bursts of transmission followed by idle periods. A data network should be able to take advantage of the idle periods in one application by using the network links to carry traffic from other applications that happen to be active at that time.

Effect of Delays

Many data applications can tolerate short random delays of up to one second. Random delays would, however, distort speech beyond recognition. A data network can take advantage of this to enable it to store data for short periods until transmission resources become free.

7.3.3 Packet-, Frame-, and Cell-Switched Networks

Efficiencies can be achieved in networking bursty data if the data is divided into blocks called packets, frames, or cells. Packets, frames, and cells have no strict definition, but they are generally understood as follows:

- Packets are usually small, variable length units of data;
- Frames are usually large, variable length units of data;
- Cells are small, fixed length units of data.

In the following discussion we talk about all three data types; however, any given network will only use one of the three types.

In packet-/frame-/cell-switched networks, the calls only occupy the network circuits when data is being transmitted. In this way, many applications can share the same circuits in the network on a time-sharing basis (Figure 7.3). Examples of technologies include:

- Packet switching: X.25 and TCP/IP (the Internet);
- Frame switching: frame relay;
- Cell switching: ATM.

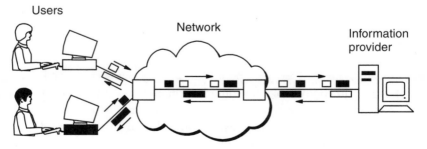

Figure 7.3 Packet-/cell-/frame-switched networks allow two or more users to access a database concurrently using the same links in the network.

7.3.4 Circuit-Switched Networks

In circuit-switched networks, calls occupy a circuit of specified bit rate through the network, regardless of whether data is being transmitted or not. You pay based on the duration of the call, regardless of how much or how little data you transmit or receive. In circuit-switched networks, a separate circuit must be established through the network for each communication (Figure 7.4).

The PSTN is a good example of a circuit-switched network, but it is not a data network as such. The best example of a circuit-switched data network technology is ISDN, which is designed to carry both voice and data.

A small number of countries, including Germany, Austria, Scandinavia, and Japan, have a public circuit-switched data network according to the X21 standard. Given the lack of international use, this type of network is not particularly popular. It has been discontinued in France.

7.4 Understanding Data Networks

7.4.1 Logical Channels

The connection between the customer's premises and a data network is known as the *access line.* In most cases this line can have multiple communications occurring at the same time. Thus, we say the access line is split into *channels.*

On circuit-switched networks, the channels have a fixed bit rate. For example, a basic-rate ISDN access line consists of two 64-Kbps channels and one 16-Kbps channel.

On packet-switched networks, the channel allocation is much more flexible. You can have as many channels as you want (within practical limits), and they have a variable bit rate up to the bit rate of the access line (Figure 7.5).

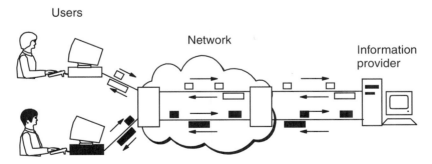

Figure 7.4 In circuit-switched networks a separate circuit is established through the network for each communication.

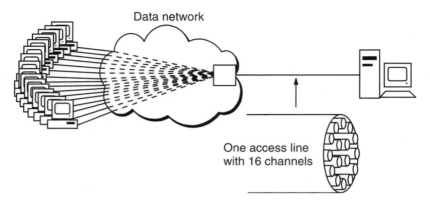

Figure 7.5 An access line can be split up into many logical channels allowing many communication links to be set up concurrently.

In packet-switched networks, these channels are known as *logical channels*. For example, a 64-Kbps access line could have 16 logical channels, each one having a maximum bit rate of 64 Kbps and all sharing the access link on a time-sharing basis. This would allow connection to 16 other computers at one time over the one access link.

It must be pointed out that only one logical channel can use the full 64 Kbps a time. If all 16 channels were active at the same time, average throughput for each channel would reduce to 4 Kbps. On the other hand, if the traffic was very bursty, then the chances of more than two or three channels being active at one time might be very small and each logical channel would achieve a much higher throughput for each burst of data.

The principle is used extensively on the Internet to give multiple users concurrent access to particular web sites.

7.4.2 Virtual Circuits

In packet-/cell-/frame-switched networks, a connection across the network is called a virtual circuit. The word *virtual* hints at how the communication path does not use up any capacity unless there is actual data being transmitted.

In packet-/cell-/frame-switching jargon, two types of connection are possible over a network: a *switched virtual circuit* (SVC) and a *permanent virtual circuit* (PVC).

Switched Virtual Circuits

Switched virtual circuits are, in some ways, similar to the type of circuit we get every time we use a telephone. It is set up when we want it and broken down when we are finished.

Permanent Virtual Circuits

Permanent virtual circuits are like leased lines. They can remain connected for months, years, or even decades. This may sound like a waste of resources, particularly at times when no information is being sent. However permanent (or indeed switched) virtual circuits established over a packet-/cell-/frame-switched network do not use any transmission capacity when no data is being transmitted.

PVCs are established by the network owner and not by the customer. They have the following advantages over SVCs:

- There is no call setup delay when you want to send data.
- Computers connected via PVCs are less vulnerable to hacking because they cannot be dialed up.

PVCs may be the only choice in new network technologies (e.g., public frame-relay providers only offered PVCs up to 1995 and many still do not provide them in 1997). The standards for frame-relay PVCs were available in 1991, whereas the standards for frame-relay SVCs have only been available since January 1994. A similar pattern is emerging with ATM. This is because the standards for PVCs are easier to define and are more fundamental to a connection-oriented data networking technology. They are therefore defined and developed first.

The big disadvantage with PVCs compared to SVCs is their lack of flexibility (i.e., they do not allow any-to-any communication).

7.4.3 Communication Protocols

The term *protocol* is a fundamental term when considering data communication. A communication protocol is a set of rules governing how communication devices (network nodes, modems, computers) talk to one another. The rules are implemented in software or hardware in the communication devices. The software adds extra information to the data to be transmitted across a link to inform the other end how to handle this piece of data. This extra information is added in a consistent way according to the rules of the protocol in use.

Framing

Most communication protocols split the data into blocks called frames, packets, or cells. The protocol will normally specify that addresses, channel identification, and framing information are added to the front of the data, while checksums, used for error detection, are added to the end of the data (Figure 7.6).

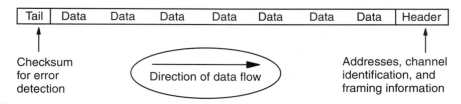

Figure 7.6 A data packet with header and checksum.

These blocks are the basic unit of data as far as error handling, routing, and interleaving of data from different virtual circuits are concerned. The network nodes, for example, use the headers to identify the packets and send them towards their correct destinations.

Protocol Stacks

There are many different data communication protocols in use today. Protocols usually work together in a group referred to as a protocol stack. For example, the Internet uses a protocol stack called TCP/IP. TCP/IP is also used on UNIX networks, and it is becoming popular on PC LANs. IBM developed the *system network architecture* (SNA) protocol stack for their mainframe networks. The protocol stack used to access packet-switched networks is called X.25.

Functions of Protocols

Protocols handle many functions. Among them we can include:

- Sending of address information;
- Sending channel identification numbers;
- Encryption of data;
- Compression of data;
- Framing (i.e., splitting data into frames, packets, or cells);
- Repeat attempts and time outs;
- Error detection and correction.

Layering of Functions

At an early stage in the development of data-transmission techniques, it was realized that it is best to organize these functions into separate *layers,* with each layer responsible for a fixed set of functions. In this way, it would be possible to interchange equivalent layers from different protocol stacks.

Open Systems Interconnect Model

The *International Standards Organization* (ISO) in the late 1970s proposed that any data communication system should have seven layers, starting at what they call the physical layer and finishing at the application layer. This model is referred to as the OSI seven-layer model. This model is neither a protocol nor a protocol stack. It is a definition of how the various functions of any protocol or protocol stack should be arranged into layers. These layers are illustrated in Figure 7.7.

Figure 7.8 shows how two popular protocols are divided into layers.

Many important data protocols were developed before the OSI model was defined and do not conform to the model. TCP/IP (the Internet protocol suite) and SNA are classic examples. All is not lost, however. Many of these original stacks are already layered in a similar way to the OSI layering, and new versions of these protocols conform more closely to the model.

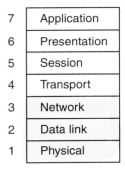

7	Application
6	Presentation
5	Session
4	Transport
3	Network
2	Data link
1	Physical

Figure 7.7 The OSI seven-layer model.

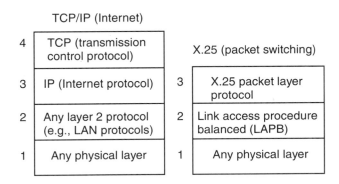

TCP/IP (Internet)

4	TCP (transmission control protocol)
3	IP (Internet protocol)
2	Any layer 2 protocol (e.g., LAN protocols)
1	Any physical layer

X.25 (packet switching)

3	X.25 packet layer protocol
2	Link access procedure balanced (LAPB)
1	Any physical layer

Figure 7.8 Some protocol stacks.

Layers Used in Data Networks

Layers 1 to 3 are all that can be found within a data network. Put another way, layers 4 to 7 are to be found exclusively on the customer's premises equipment.

7.4.4 Connection-Oriented and Connectionless Services

Communication between two devices over a network can take two forms—connection oriented and connectionless—as summarized in Table 7.1.

Connection-oriented services have a call setup phase followed by a data-transfer stage, followed by a call release phase. They are ideally suited to situations where the same two devices will be in communication for an extended period of time.

Connectionless services have no call setup phase. The full source and destination address is carried within each packet of data. Each packet is routed independently to its destination. Connectionless services are ideally suited where devices are required to communicate with a large number of different devices in quick succession because there will be no call setup delays. Because LANs use a connectionless protocol internally, it can be advantageous to link them using connectionless services.

It is possible for a network that is essentially connection oriented to support connectionless services and vice versa. Thus, for example, a connection-oriented ATM network can form the infrastructure for a connectionless SMDS service. Equally, the TCP protocol used on most end stations on the Internet provides a connection-oriented service over a connectionless network [1].

Table 7.1
Connection-Oriented Versus Connectionless Communication

Communication	Connection Oriented	Connectionless
Characteristics	Has a call setup phase Packets normally follow the same path through the network Packets arrive in order	Has no call setup phase Each packet can follow a different path through the network Packets may arrive out of order
Examples	ISDN X.25 Frame relay	Most LANs Internet protocol (IP) SMDS

7.5 X.25 Packet-Switched Networks

7.5.1 Background

X.25 packet switching is available in every developed country and in many third-world countries. The name *X.25* comes from the name of the specification of the access line protocol issued by the CCITT (now ITU-T—the standards-setting body of the International Telecommunications Union). The X.25 standard was developed in the late 1970s, and the first commercial networks were built in the early 1980s.

X.25 packet-switched networks are designed to carry bursty traffic, as described in Sections 7.3.2 and 7.3.3. You can connect to X.25 networks directly or via a dialup connection (see Figure 7.9).

7.5.2 Direct Connections

Direct connection to an X.25 network is often referred to as an X.25 connection. This is because the X.25 protocol is used between the customer's equipment and the network. Direct connections allow both outgoing and incoming calls and allow more than one call at a time over the access line.

Packet Assembler/Dissembler

For any direct connection to the network you need to have a *packet assembler/dissembler* (PAD). The PAD is the device that splits your data into packets and puts the headers and trailers on the packets. On the receive side, a PAD takes packets from the network and joins them back together so that they look like the original data stream. The PAD can either be a card installed in one of

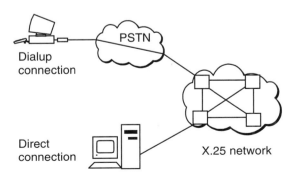

Figure 7.9 Dialup and direct connection to a packet-switched network.

your computers (Figure 7.10) or a stand-alone unit between your computer(s) and the access line (Figure 7.11). PADs cost around $800.

The PAD connects to the line via a modem in the case of an analog line or an NTU (*channel service unit/data service unit* [CSU/DSU] in the U.S.) in the case of a digital line. The modem will be a leased line modem as opposed to a dialup modem. Your network provider will inform you which type to purchase.

Stand-Alone PADs

A stand-alone PAD allows many computers to be connected to an X.25 network. The PAD has one X.25 port for connection to the network and a large number of data ports for connection to the computers.

Dialup Connections

In the case of dialup connections, you must dial a PSTN number that connects you to the packet-switched network and then enter an identity code and a password before you can start dialing other X.25 users. This identity code is called a

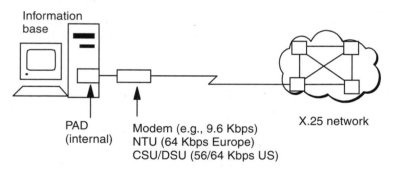

Figure 7.10 The connection of a computer to an X.25 line via an internal PAD.

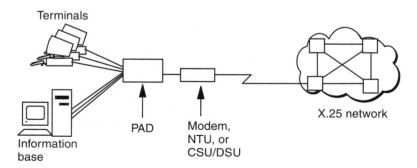

Figure 7.11 The connection of many computers to an X.25 line via a stand-alone PAD.

network user identity (NUI) and is used for billing. Dialup connections have a lower fixed cost but higher usage charges than direct connections.

There are two possible types of dialup connections. The simplest and most widely available is referred to as *X.28 dialup*. In this case, all you need is a PC, modem, and terminal-emulation software. The PAD is located in the network, and you connect to it when you dial into the network. X.28 dialup is quite a restricted service because it only allows one call at a time and does not allow incoming calls.

The second possibility is referred to as *X.32* or *dialup X.25*. In this case, the PAD is located on the customer's premises. The advantages of this is that you can make more than one call over the X.25 network at one time, and you can be called by others. Dialup X.25 is not available on all X.25 networks.

Figure 7.12 illustrates a dialup connection to a packet-switched network.

7.5.3 Connecting a LAN to an X.25 Network

LANs are connected to X.25 networks for two reasons. One is to distribute X.25 facilities to the users (Figure 7.13); the second is for LAN interconnect. The first type of connection can be achieved using a communication server. Suitable software must be installed on the communication server and on the

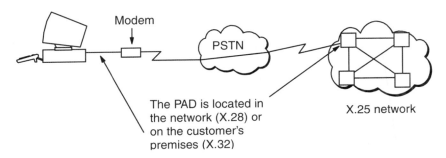

Figure 7.12 Dialup connection to a packet-switched network.

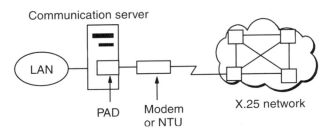

Figure 7.13 Connecting a LAN to an X.25 network to distribute X.25 facilities to the users.

workstations. This will allow any PC on the network to dial out over the X.25 network.

Interconnecting LANs via an X.25 network requires a router with an X.25 port (Figure 7.14). This type of router will have the PAD functions built in.

7.5.4 Connecting Mainframes to X.25 Networks

Mainframe computers can be connected to the end users via an X.25 network (Figure 7.15). A *front-end processor* (FEP) is connected to a cluster controller via the X.25 network. IBM has specified a protocol called *qualified logical link control* (QLLC), which makes this possible.

7.5.5 Charging

Charges on X.25 networks are split into duration and volume charges. The volume charge is based on the volume of data transmitted during a call. For long-distance and international calls, the duration charges are much lower than the equivalent PSTN charges. The combination of duration and volume charges normally works out to be less than the equivalent PSTN charge when using interactive applications where some of the time is spent reading the screen or deciding what to do next. File transfers, on the other hand, can work out to be more expensive than using the PSTN. Figure 7.16 provides a comparison between PSTN and X.25 call charges.

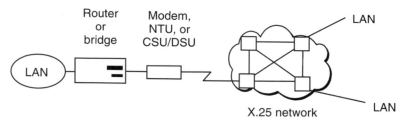

Figure 7.14 Interconnecting LANs via an X.25 network.

Figure 7.15 Connecting terminals to a mainframe via an X.25 network.

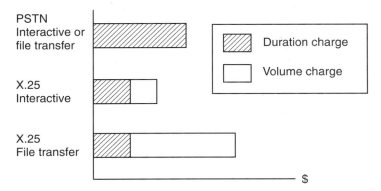

Figure 7.16 Comparison between PSTN and X.25 call charges.

7.5.6 Typical Uses of X.25 Networks

X.25 networks are used to provide wide area access to a text-based information service (e.g., terminal access to a mini or mainframe computer). In many cases, it is significantly cheaper to dial via the X.25 network than to dial via the PSTN. It can also work out to be cheaper to use dedicated X.25 connections rather than leased lines to permanently connect branch offices to the head office computer. The cost differentials tend to increase with distance. Many businesses use X.25 to allow their employees in branch offices or on the road to gain access to company information.

X.25 networks can also be used to interconnect LANs. X.25 has the advantage of widespread availability and the security offered by closed user groups or PVCs. X.25 will present few problems if the only application is email between the LANs. Client/server applications should also work well, provided they do not generate large data flows. It is possible to transfer delay sensitive traffic such as SNA across an X.25 network. However, special care is required in configuring the customer's equipment to take account of the network delays. Transfer times for medium to large file transfers will be poor as a consequence of the relatively low access speed and the large delays in the X.25 network.

X.25 networks are normally used to provide the network infrastructure for X.400 email systems (see Section 10.3.1).

7.5.7 Other Features and Facilities of X.25 Networks

X.25 networks around the world are interlinked to form a global network. This is a considerable advantage over some of the newer networking services, such as

SMDS and ATM, which are not fully interlinked into a global network at present.

Reverse charge calls are also possible. This makes it possible for information providers to invoice you with a single bill covering both your X.25 charges and your information retrieval charges.

7.6 Frame Relay

7.6.1 Background

Frame relay is a data-networking technology that allows both switched and permanent calls at higher throughput than X.25 networks. Users of a frame-relay service are connected at speeds ranging from 56 Kbps up to 2 Mbps. Some manufacturers of frame-relay equipment have developed access speeds of up to 45 Mbps.

Because only PVCs are on offer in many parts of the world, frame relay is currently viewed as an alternative to leased lines. Depending on the user requirements and the tariffs, frame relay can work out to be anywhere between 20% and 70% cheaper than leased lines. The advent of SVC services should also make frame relay a viable alternative to ISDN or X.25 for small branch offices.

The primary use for frame-relay services is LAN interconnect, email, and Internet access. Voice and video over frame relay are possible but are less widely used. Frame relay also has the facility to carry IBM's SNA protocol quite efficiently and can thus be used to connect the various components in an IBM mainframe network. There is a growing interest in the use of frame relay as a means to integrate LAN interconnect and SNA traffic on the one access line.

7.6.2 Basic Concepts

Faster Than X.25

Frame relay is a simpler and thus faster protocol than X.25, making it better suited to bandwidth-hungry applications, particularly LAN interconnect. Frame-relay networks have no internal error correction. This is what makes the protocol simple and fast. Error rates are kept low by using digital transmission links. Any errors that do occur can be handled by the terminal equipment (routers, PCs, minicomputers) (see Figure 7.17(b)).

One drawback of leaving error correction to the terminal equipment is that significant disruption is caused when error correction does occur. In most cases, all of the data transmitted since the error must be retransmitted. On high-speed connections, this can amount to a significant volume of data. Some appli-

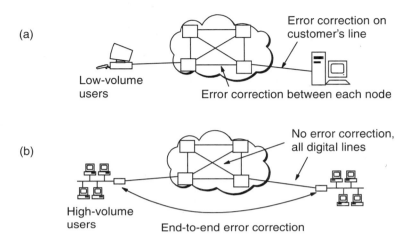

Figure 7.17 (a) X.25 and (b) frame relay compared.

cations such as real-time voice and video allow errors to pass uncorrected, as the recovery would cause greater degradation than the error itself.

Unsuited to Analog Networks

As already noted, frame-relay networks must use digital links. This requirement applies equally to the access lines and the internodal trunks. For this reason, frame relay will not be available for some time in less developed countries, which still rely on analog transmission systems. X.25 will remain the preferred option for public data networks in these countries and is thus likely to be with us for some time to come.

Committed Information Rate

Each virtual circuit on a frame-relay network (i.e., PVC or SVC) has an associated *committed information rate* (CIR). This is the bit rate that the network will guarantee for that connection. PVCs have a fixed CIR that can only be changed by the network provider. The CIR for SVCs can be negotiated during call setup.

Excess Information

The CIR for a PVC can be, and often is, set below the speed of the access link. Thus, for example, a PVC may have a CIR of 32 Kbps, but an access line speed of 64 Kbps. If there is no congestion in the network, the *customer's premises equipment* (CPE) may be allowed to transmit at the full line speed of 64 Kbps. When this happens, half of the frames transmitted will be marked *discard eligible.* Discard-eligible frames are the first to be discarded if the network experiences congestion.

The network can inform the CPE that there is congestion in the network. Good CPE will slow down their transmission to at least the CIR to ensure that no information is marked discard eligible during the period of congestion. If this happens quickly enough, there is a good chance that no frames will be discarded. Poor CPE, on the other hand, will continue to transmit above the CIR and will suffer the consequences; frames will be discarded, resulting in retransmissions and even dropped sessions (where an application has to wait too long for a response and the user's screen locks up).

Excess Information Rate

Excess information rate (EIR) is the rate above the CIR at which the equipment is allowed to send discard-eligible data. It is normally the difference between the CIR and the access line speed. Some network operators allow the CIR + EIR to be less than the line speed. In this case, the customer must be careful to ensure that his or her equipment can be programmed never to exceed CIR + EIR on a PVC. Otherwise, the network will definitely discard frames, even causing problems under normal conditions. Figure 7.18 illustrates the allocation of bandwidth for a typical PVC on a 64-Kbps frame-relay access line.

Bursting

Frame relay allows traffic to burst above CIR + EIR for short periods of a second or two, provided that the *average* transmission rate does not exceed CIR + EIR. These bursts are allowed if the PVC has been idle for a period before the burst. Bursting helps to improve the response time of interactive (bursty) applications.

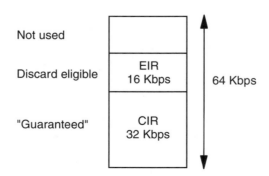

Figure 7.18 The allocation of bandwidth for a typical PVC on a 64-Kbps frame-relay access line.

Zero CIR

Some network operators offer a zero CIR service (i.e., all traffic is sent discard eligible). They offer attractive rates for this service. If they have plenty of capacity in their networks, very few problems will be encountered. There is always the danger, however, that if congestion becomes a problem, you have no guarantees about delivery of your data. A way around this problem is to have guarantees about the quality of service stated in your contract.

7.6.3 Typical Setup

Frame-relay networks are used primarily for LAN interconnect. This is normally achieved using a router or bridge with a frame-relay *user network interface* (UNI), which is connected to the frame-relay access line (Figure 7.19). Many routers manufactured since 1993 will be upgradeable to frame relay.

A *frame-relay access device* (FRAD) allows various inputs to be multiplexed onto a frame-relay access line (Figure 7.20). Most FRADs have an input port for a LAN and some combination of additional ports for terminals or X.25 equipment. Some FRADs can carry voice traffic. The type of input allowed will depend on the particular model of FRAD.

Mainframe manufacturers (e.g., IBM) have recognized the importance of frame relay and have implemented frame-relay UNI in their FEPs (Figure 7.21).

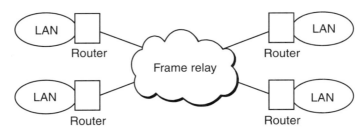

Figure 7.19 LANs interconnected by routers over a frame-relay network.

Figure 7.20 FRADs allow LANs and other equipment be connected over frame relay.

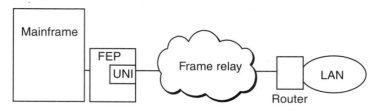

Figure 7.21 FEP connected to a frame-relay network.

Some things to look for in a FRAD or router are:

- *Rate control:* the ability never to exceed CIR + EIR for a sustained period;
- *Discard-eligible bit control:* the ability to set the discard-eligible bit selectively on low-priority traffic;
- *Congestion control:* the ability to throttle back to the CIR in the presence of congestion.

7.6.4 Sample Application and Cost Calculation

A company wishes to interconnect LANs in 12 locations. All sites require email facilities. Seven sites require continuous interconnection to allow client/server applications to work between them. The majority of the servers are in the headquarters site.

The solution chosen is to interconnect the seven sites using routers connected by frame-relay PVCs. The frame-relay connections are cheaper than using six leased lines while giving similar performance for client/server applications. If possible, the routers would be programmed to send all client/server traffic as CIR traffic (i.e., guaranteed delivery). The email traffic, on the other hand, would be sent discard eligible.

This setup will reserve the CIR bandwidth for the time-critical client/server applications and let the email do the best it can. In practice, however, this could prove difficult or impossible. It depends on the capabilities of the routers and the time and expertise at the disposal of the network manager.

The other five sites do not require online connectivity and will use dialup ISDN bridges to connect to the headquarters every 30 minutes to send and receive email. These calls will typically remain connected for less than 10 seconds each time. (Average email messages are 1K—more than 60 messages could be transferred in 10 seconds. Alternatively a 20-page document could be transmitted in 10 seconds.)

Thus, each site would make about 20 ISDN calls per day equal to 200 call seconds or 3.3 call minutes. In most cases, the cost of the 20 calls would be 20 times the minimum rates for an ISDN call, about $2 per day or $520 per year. Adding a year's rental for an ISDN line gives us a total of about $1,200 per year per site (ballpark figures, 1996). Frame-relay PVC tariffs cannot compete with this use of ISDN.

7.7 Switched Multimegabit Digital Service

7.7.1 Background

SMDS is a public data service designed primarily for the interconnection of LANs at high speeds. Unlike frame relay, SMDS began life as an any-to-any service. That means that once you are connected to an SMDS network, you can make calls to any other connected user within the constraints of security arrangements, such as closed user groups. SMDS is a connectionless service. This further enhances its suitability for LAN interconnect. Until switched ATM services become available, SMDS is the only LAN-interconnect service providing any-to-any communication and bandwidth on demand with access speeds above 2 Mbps.

SMDS Speeds

SMDS allows users to be interconnected at speeds in the range 2 to 34 Mbps in Europe or 1.5 to 45 Mbps in the United States. Some service providers offer low-speed SMDS services starting at 56 or 64 Kbps. Low-speed access increases the scope for connecting small branch offices. The backbone links in the network work at speeds of up to 155 Mbps at present.

SMDS has a high overhead. For example, a T1 access line operating at a rate of 1.5 Mbps can only support a user data rate of 1.17 Mbps. The rest of the bandwidth is overhead (i.e., protocol headers and checksums). Low-speed access links (64 and 56 Kbps) use a different protocol and do not suffer from this overhead problem (see Section 7.7.3).

SMDS and MANs

As with X.25 and frame relay, SMDS does not define the network infrastructure; it only defines the user's interface to the network. Most SMDS networks today are based on the IEEE 802.6 *metropolitan area network* (MAN) standard. A MAN is a network within a city. MANs are very often implemented in a ring topology, as in this configuration they can reconfigure and still work at full capacity in the presence of a cable break. MANs can be interconnected, and there is no ultimate geographical limit to SMDS services based on MAN technology.

SMDS and ATM

SMDS services can also be offered on ATM networks. At present there are already similarities between MANs and ATM; for example, MANs and ATM networks both use 53-byte fixed-length cells to carry data.

Security

A public SMDS service is shared by many commercial users who wish to interconnect their own LANs but do not want other network users to have access to their LANs. Safeguards against security breaches include authorization/address screening (at source and destination) and the provision of *virtual private networks* (VPNs).

Other SMDS Characteristics

Other SMDS characteristics include:

- CCITT E.164 addressing—SMDS addresses are like standard telephone numbers;
- Support for main protocols (e.g., TCP/IP, Novell, DECNet, Apple-Talk, SNA, OSI);
- Multicast broadcasting capability, an advantage for LAN interconnection.

7.7.2 Applications

The primary application for SMDS, especially in the initial stages, is LAN interconnection. By using SMDS, users accessing a service on a remote LAN will get similar performance to local users.

In general, SMDS can be used for any application where large amounts of information must be cost effectively transferred at high speed. Typical examples include:

- Medical imaging;
- Financial and office image transfer;
- Design graphics;
- Multimedia computing;

- Print publishing;
- Distributed computing;
- Networking of high-performance workstations.

7.7.3 Connection to an SMDS Network

LANs are connected to SMDS networks via a router and an SMDS-compatible CSU/DSU. The SMDS-compatible CSU/DSU is responsible for segmenting the data into cells and as such is significantly more complex than a standard CSU/DSU. The interface between the router and the SMDS CSU/DSU is called the SMDS *data exchange interface* (DXI).

Low-Speed SMDS

Some SMDS providers supply access to their networks at 56 or 64 Kbps. This option is of obvious interest to small businesses and small branch offices. To reduce the cost of such connections, standard CSU/DSUs are used on the customer's premises. The data is not segmented into cells until it reaches the first node in the SMDS network. This has the added advantage of making maximum use of the limited bandwidth on the low-speed access line because it is the process of splitting the data into cells that creates the large overhead associated with SMDS. The difference between high- and low-speed connections are illustrated in Figure 7.22.

7.7.4 SMDS Versus Frame Relay

SMDS, being connectionless and providing any-to-any connections, has certain advantages over frame relay, particularly for networks with many sites. For example, a U.S. company with 54 sites to interconnect analyzed three different

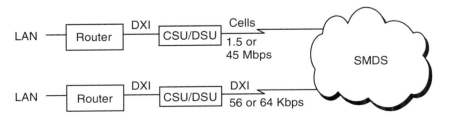

Figure 7.22 High- and low-speed connection of a LAN to an SMDS network compared.

network architectures in 1994: SMDS, public frame relay, and private frame relay. It selected SMDS because of:

- "The simplicity of its connectionless protocol;
- Its peer-to-peer networking capabilities, as well as multicast broadcasting capability;
- Upward scalability, which makes it easy to add sites without reconfiguring the network;
- Price/performance" [2].

The company's choice of SMDS highlights the advantage of any-to-any communications in a network with a large number of nodes. At the time of implementation (late 1994), frame relay could only offer PVC connections. If the company had gone down the frame-relay road they could have ended up with over 100 PVCs to manage. The solution chosen depended on the availability of low-speed (56-Kbps) access to SMDS for the branch offices. Low-speed service is not available on all SMDS networks.

7.8 Asynchronous Transfer Mode

7.8.1 Background

ATM is a data-networking technology that has only recently emerged from the development phase. It is designed to carry all traffic types (voice, video, and different types of data) efficiently and with better guarantees about service quality than that offered by any other network discussed so far. ATM's ability to carry all traffic types should give it an economic advantage over other technologies. ATM is a broadband technology (i.e., it is designed to transport and switch traffic with very high bit rates). ATM has been chosen by the ITU as the platform for the next phase of ISDN, known as *broadband ISDN* (B-ISDN).

The following contrasting traffic types can all be carried efficiently by an ATM network:

- High and low bit rate;
- Variable and constant bit rate;
- Delay sensitive and nondelay sensitive;
- Connectionless and connection oriented.

The first services to become available over public ATM networks were PVCs (1995). Switched services are expected to emerge soon.

7.8.2 Network Operation

Data is carried in fixed-length cells 53 bytes long (Figure 7.23). Each cell has a five-byte header.

The header is used for routing and congestion control (Figure 7.24). There is no trailer to the cells. The header contains a checksum to ensure no cells are routed in error. All error correction of user data is left up to the terminal equipment. Due to the fixed length cells, ATM is sometimes referred to as *cell relay*.

One of the drawbacks of ATM is the high overhead in the header. The header does not contain any user data but takes up approximately 10% of the bandwidth. This high overhead is a direct result of using small cells. In addition, many end-user protocols further increase the overhead associated with ATM.

ATM Versus Frame Relay

Like frame relay, ATM relies on quality digital transmission links throughout the network and leaves error correction to the terminal equipment. Unlike frame relay, ATM uses short fixed-length cells. This helps to further simplify the switching process and permits higher speed switching with lower delays. Thus, ATM is better suited to voice, video, and very-high-bandwidth applica-

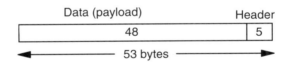

Figure 7.23 An ATM cell.

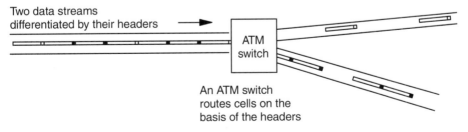

Figure 7.24 ATM switching.

tions. Frame relay, on the other hand, has a head start on ATM, so there are more products and services available.

It is expected that many ATM networks will allow customers to connect frame-relay devices to the network. The conversion from frame-relay format to ATM will be carried out at the interface to the ATM network. This interface is called *frame-relay user-network interface* (FUNI). This will mean that investment in frame-relay equipment will not be wasted. This has been an important factor in convincing network managers to invest in frame relay rather than waiting until ATM arrives.

There is a school of thought that says that frame relay will be the preferred method of connecting low-speed sites to ATM networks. The reason is that frame relay has much lower overhead than ATM, and while this overhead may be tolerated on high-speed ATM backbones, it will not be welcome on low-speed links to branch offices where every ounce of bandwidth is precious.

ATM permits switching at speeds an order of magnitude greater than frame relay, making it more suitable for very-high-speed backbone networks.

Traffic Types

Figure 7.25 shows the various types of traffic that ATM attempts to carry.

The *available bit rate* (ABR) service is particularly important, as it allows you to get the most out of an ATM network. The idea is to give nonurgent data access to any network capacity not being used by more urgent applications.

ATM LANs

ATM technology is already in use in large LANs. It can be used as a backbone technology in place of FDDI or as a complete network technology delivering ATM to the desktop. In most cases, ATM to the desktop will be mixed with conventional LANs. The ATM equipment is installed in areas that require the bandwidth or multimedia capabilities.

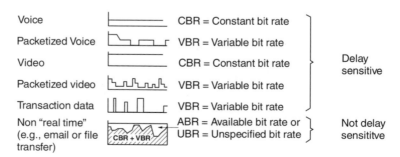

Figure 7.25 Some of the main types of traffic that ATM networks must carry.

One advantage of installing an ATM backbone is that one or more of the backbone switches can be easily connected to a public ATM network.

ATM PBXs

Most of the development effort in ATM so far have been directed at the problems of LAN interconnection and multimedia. It is envisaged that ATM will also be used to carry standard telephony services. A first step in this development is an interface to connect a PBX to an ATM switch. Such an interface should allow LANs and PBXs to share capacity on an access link into an ATM network such as that illustrated in Figure 7.26.

Call Establishment

Call establishment requires information such as the destination address to be sent to the network. This information is sent on a separate signaling channel in much the same way as the D-channel is used for signaling in ISDN. Once the call is established, a different channel is allocated for the data transfer.

During call establishment, the calling equipment will inform the network of the class of service it requires. For example a voice call would require CBR service with a specified maximum delay variation. This requirement would be signaled to the network. The network would then check if it has the resources (spare capacity) to establish that call. If it does, the call is established; if not, the user equipment is informed.

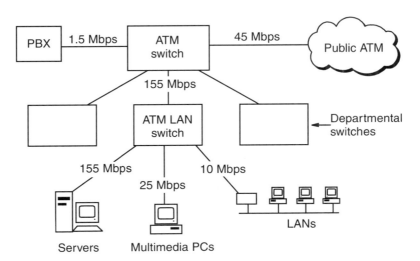

Figure 7.26 A large campus ATM network.

7.9 VSAT Networks

A *very small aperture terminal* (VSAT) is a small satellite dish, and associated electronics, pointed at a geostationary satellite (see Section 2.3.2). The main advantages of VSAT networks are that they are often the lowest cost solution, particularly when there are a lot of sites and the distance between the sites is large. In addition, satellite-based communication is often more reliable than leased lines and is more feasible in remote areas.

On the negative side, the satellite hop introduces a 250-ms end-to-end delay in the transmission path, which makes it unsuitable for some delay-sensitive traffic. Satellite communication is relatively easy to eavesdrop, so encryption will be required for sensitive communication. In some countries, satellite communication is subject to restrictive regulations, which can make its deployment difficult and expensive.

7.9.1 Types of VSAT Networks

Three modes of operation for a VSAT network are outlined in Table 7.2.

The most common topology for a VSAT network is a star network, with everything passing through a central hub site (Figure 7.27). The hub contains the switching equipment and has a higher capacity satellite dish than the remote sites. The hub site can be owned by the satellite organization and shared by a number of organizations, or it can be privately owned and dedicated to one organization's network. Private hubs become economic when the number of VSAT terminals is large.

Table 7.2
Modes of Operation for VSAT Networks

Name	Equivalent to	Typical Use
Time-division multiple access (TDMA)	Packet switching (e.g., X25)—bandwidth is only assigned for actual data	Database access
Demand-assigned multiple access (DAMA)	Circuit switching—a fixed transmission capacity is assigned for each call	Voice or file transfer
Single channel per carrier (SCPC)	Private circuit—a fixed bandwidth is assigned between two sites 24 hours per day	Leased-line alternative

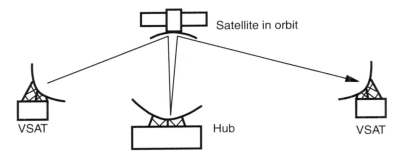

Figure 7.27 In a star network, communication between two remote VSATs passes through a hub.

Sending traffic through the hub doubles the delay when communicating between one remote site and another. The total roundtrip delay will be greater than one second, which may be unacceptable in some instances. To overcome this problem, a mesh network can be built. In this case, the hub is used during call establishment to allocate bandwidth, but once the call is established the VSATs communicate directly via the satellite.

7.10 Security of Public Data Networks

Security breaches on the Internet and on Internet hosts are openly discussed on the Internet and even in the popular press—making the Internet synonymous with insecurity. On the other hand, security breaches involving managed networks are usually kept quiet because either the owners of the network or the users (e.g., financial institutions) have much more to lose by the report of an incident than the incident itself.

Many managed networks have good additional security features, such as closed user groups and PVCs, but these alone may not be sufficient in all circumstances. The professional hacker whose aim is to make money from his or her activities will be attracted by the lucrative targets attached to these managed networks. Techniques such as physical wire tapping, interception of microwave radio signals, or even a physical break in to your premises or the network provider's location will render many security features ineffective. A particular risk with managed networks exists if the hacker gains control of the network-management system. The physical and logical security procedures associated with a network-management system is of critical importance, regardless of whether it is your own private system or a system controlled by a network operator.

If your data is sensitive or valuable, then additional security should be added through the use of encryption and by strengthening the access security required for logging onto your host computer systems. Management and administration are as important as the security hardware and software used to protect your systems. If the hacker can get his or her hands on encryption keys or passwords, then you might as well not have a security system.

References

[1] Piscitello, David M., and A. Lyman Chapin, *Open System Networking—TCP/IP and OSI*, Reading, MA: Addison-Wesley, 1993.

[2] SMDS Interest Group, "A look at Downey Savings' network," http://www.smds-ig.org, 1994.

Further Reading

Brewster, R. L. (ed.), *Data Communications and Networks 3,* London: Institution of Electrical Engineers, 1994.

Hopkins, Harmon, *The Frame Relay Guide: The Handbook of Frame Relay and Fast Packet Techniques,* London: CommEd, 1993.

These books provided some of the ideas used in this chapter.

8

The Internet

8.1 Background

The Internet is the world's largest and best known computer network. It is often referred to simply as *the net*. It spans the globe and is readily accessible in all developed countries. Access in the developing world is, however, quite poor, particularly in large portions of the African continent.

The Internet evolved out of a number of separate networks, notably the *advanced research project agency network* (Arpanet), a research network that grew up in the 1970s. In the 1970s and 1980s, the principal users of the Internet were universities and the U.S. Department of Defense. In recent times, the Internet has been commercialized. Commercial organizations use it for promotion of their products, sending email, and delivering technical support.

A very important factor in driving the commercialization of the Internet was the development of a user-friendly interface (the web browser) in 1992. Not only are web browsers easy to use, they allow the same information to be viewed from a variety of computer types (i.e., PCs, Apple Macintoshes, and UNIX workstations).

In the period 1993 to 1997, the main commercial activity on the net has been the promotion of products and services and the use of email. There is great interest in electronic cash and secure submission of credit card numbers over the Internet. The take-up of this type of service has been low to date; however, this situation is expected to change radically over the next few years.

Estimates of the number of computers with Internet access varied from 35 to 100 million in early 1997. The exact number is considered impossible to determine because an organization only needs one connection to the Internet to give access to all computers on its internal network. There are also differences surrounding what exactly constitutes Internet access. Lower figures will be obtained if you include only computers with World Wide Web access, while a much greater figure will be obtained if you attempt to estimate the number of people with Internet-style email addresses.

8.2 Connecting to the Internet

Today anybody with a computer, Internet software, a modem, and a phone line can connect to the Internet by paying a fee to a local Internet service provider (ISP). It helps if the ISP has a *point of presence* (POP) close by because this keeps the call charges down. Large organizations find it more economical to have a dedicated communication link to their ISP.

Internet software can be subdivided into communication software and application software. The communication software includes a TCP/IP protocol stack—the protocol stack used on the Internet (see Sections 7.4.3 and 12.3.2). The TCP/IP protocol stack is built into many desktop operating systems (such as UNIX or Windows 95).

There are many Internet applications; most common are the web browser, email software, telnet software, a newsreader, and phone software. Much of this software is either low cost or free.

8.2.1 The Single User Connection

A dialup connection over the PSTN or ISDN is an economic way for a single user or small business to connect to the Internet (Figure 8.1). The minimum requirements are a modem, a telephone line, and a terminal (usually a PC).

ISDN access is supported by many ISPs. Its advantages over the PSTN are twofold: first, you can connect in about 4 to 10 seconds, compared to 20 to 30 seconds using a modem; and second, your connection will have two to four times more bandwidth. In the case of ISDN, a terminal adapter is used in place of the modem.

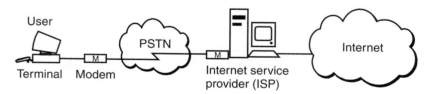

Figure 8.1 A single user dialup connection to the Internet.

The software at the user's end can be configured to dial up periodically and check for email. Alternatively the ISP may offer a service whereby it dials into your computer whenever you have mail waiting. You, of course, have to pay for these calls. It is important to ensure that some security mechanism is also offered to avoid someone else dialing into your computer.

Some ISPs offer a dedicated modem service. This is a modem on the ISP's end reserved for your connection, meaning that you will never get busy tone when you dial in.

People who travel a lot can save on international call charges by subscribing to an ISP that has POPs in each country they visit. In this way, they will be able to keep up to date with their email without having to make international phone calls. Large multinational ISPs and some smaller ISPs, which have formed alliances, can offer this type of service.

8.2.2 Corporate Connections to the Internet

Large businesses can connect their LANs to the Internet via dialup PSTN or ISDN, frame relay, or leased lines. They can thereby give everybody in the organization access to the Internet. They can restrict access to certain applications while allowing access to Internet email (browsing the Internet is often considered to be a waste of time for particular employees).

A router is required to connect to the ISP, and a firewall (see Section 8.6.1) is required to protect your corporate data against hackers. Many ISPs offer *managed* routers on the customer's premises and firewall services. This is particularly attractive for smaller organizations that do not have the technical expertise to purchase and manage their own router or firewall. A typical setup is shown in Figure 8.2.

The decision as to which access services to use (PSTN, ISDN, frame relay, leased line) will depend on economics, traffic levels, and whether or not you will

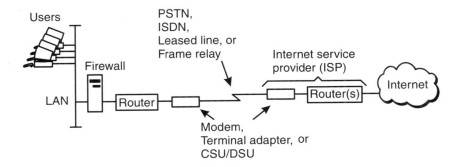

Figure 8.2 Connecting a LAN to the Internet.

be hosting information from your end of the connection. *Hosting* means setting up a server with online information and possibly the means to order products online. It is not necessary to host information from your end of the connection. It is often cheaper and better to pay your ISP to host your web site on one of its servers (see Section 8.4.1).

Permanent Connections to the Internet

If you do intend to host, then a fast, 24-hour-a-day connection is required. ISDN and PSTN connections tend to be expensive under these conditions, although there are exceptions—where ISDN or PSTN are offered at a flat rate for local calls.

Leased lines are probably the most common means of providing high-bandwidth permanent connections to an ISP. They are likely to be the only option available if you require an access speed greater than 1.5 Mbps.

Frame relay will often work out to be more cost effective, particularly if you are already using it for other purposes. Consider, for example, a business that has interconnected its LANs using frame relay. It has already invested in the frame-relay access at each site. To add a connection to its ISP, all that is required is an additional PVC from one of its sites through the frame-relay network to the ISP (see Figure 8.3). The charge for this additional PVC will be a fraction of the cost of a leased line. Care will need to be taken, however, that there is sufficient bandwidth in the frame-relay access link at the particular site connected to the ISP.

If you are not hosting information, then you will probably only require a connection during working hours and possibly only for a small number of hours per day. In addition, if you do not have too many employees browsing the web, then your bandwidth requirements will likely be modest. In these situations, ISDN and even PSTN connections should be given consideration. ISDN in particular is a very likely candidate for organizations with under 100

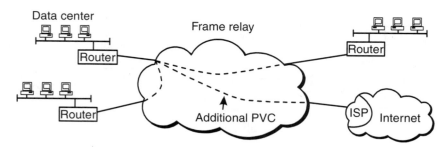

Figure 8.3 Adding an additional PVC to connect to an ISP over frame relay.

users—particularly if email is the primary application. A lot will depend on the amount of usage and on how ISDN tariffs compare to frame relay and leased lines.

8.3 Network Operation

8.3.1 A Network of Networks

Technically speaking, the Internet is an internetwork (or internet with a small "i")—a network of networks, or a network of subnetworks. In most cases the subnetworks (subnets) are LANs or groups of interconnected LANs. The subnets are linked using routers. As the name suggests, routers are used to send the data packets in the right direction.

Traffic handling on the Internet is based on the TCP/IP protocol suite. These protocols are responsible for routing traffic, avoiding congestion, and correcting errors. TCP/IP is a packet-switching protocol ideally suited to bursty applications. It differs from typical X.25 implementations (see Section 7.5) in that it allows larger packets. This makes it better suited to the interconnection of LANs. Internetworks, routers, and TCP/IP are dealt with in more detail in Chapter 12.

8.3.2 ISPs and Backbones

The commercialization of the Internet during the 1990s has radically changed the structure and ownership of the Internet. Once a network with large government sponsorship, the Internet is now dominated by ISPs and multinational giants such as AT&T, MCI, and Sprint. These companies are ISPs in their own right but also act as service providers to the smaller ISPs.

Large ISPs have high-speed backbone networks linking their various POPs [1]. Backbone bandwidth can be anything from 1.5 Mbps up to 622 Mbps (MCI, early 1997). ISPs have to link their networks together. In the United States, the biggest interconnection points are called *network access points* (NAPs). There are similar interconnection sites around the globe (but they are not always called NAPs).

The interconnection between ISPs plays a very important role in producing an efficient and robust Internet. Unfortunately, however, commercial and competitive pressures can result in inefficient arrangements. For example, at the time of this writing, in Ireland not all service providers use the same interchange point. Thus, a proportion of traffic that should stay within the country actually gets to its destination via the United States or mainland Europe.

8.3.3 Bandwidth

Bandwidth on the Internet is a problem for users and ISPs alike. This is because Internet traffic is growing so fast that any increase in bandwidth quickly becomes saturated by an increase in traffic. MCI, for example, experienced traffic growth of 30% per month during 1996. In Europe, it is quite noticeable that the Internet slows down in the afternoon when users in the United States start to go online.

The important parameter is end-to-end bandwidth, not the bandwidth of individual links. Congestion from other Internet users is a critical factor here. End-to-end bandwidth is governed by the weakest link between you and the other end. If you dial up with a 14.4-Kbps modem, this will often be the limiting factor. If, however, a particular site is very busy, no increase in the bandwidth at your end is going to compensate for the congestion at the far end. The same is true when there is congestion on the backbone.

Response time from a particular web site can also be adversely affected by the web server at that site. If a lot of people are accessing a server at a given time, the server can become a greater bottleneck than the Internet connection.

The Internet does not give any quality of service guarantees, but individual ISPs may be able to offer certain guarantees, provided your traffic remains within their network. This principal can be further extended to ISPs which cooperate closely with one another. In the future, new technologies and protocols, such as *Internet protocol version 6* (IPv6) (see Section 12.3.3), will assist ISPs in offering *grade of service* (GOS) guarantees.

8.3.4 Other IP Networks

Some service providers offer connections to an IP network that is separate from the Internet. These networks offer higher end-to-end bandwidth and lower security risks. They use the same Internet software on the user's computers. Sprint Corporation, for example, offers such a service and through the use of gateways allows its customers the ability to connect to either this network or the Internet over the same access line.

8.4 Internet Applications

8.4.1 World Wide Web

The WWW or the *web* is a name used to describe the Internet as seen through web-browser software, such as Netscape Navigator or Microsoft Internet Explorer—the two most popular web browsers. Viewed this way, the Internet looks like a web of interlinked documents or pages. There is more to the Internet than the web; however, more and more Internet services are becoming accessible via web browsers, and the distinction between the web and the Internet is diminishing.

A web page is a *hypertext* document containing built-in links to other documents. Alternatively, the links can be to downloadable files, sound or video clips, email addresses, or Internet phone connections. Web pages are stored on servers known as *web servers*. Forms can be included in a web page to allow the user to send information to the web server (Figure 8.4). This information could be an order for equipment, a response to a survey, a request for technical support, or the criteria for querying a database.

WWW Forms and Common Gateway Interface

When a user submits a form, information is transmitted back to the server and the server executes a *common gateway interface* (CGI) script using this information as input data. The server then delivers an appropriate response to the user. In the case of a database query, for example, the server will generate a web page on the fly containing the search results. The server may also perform some actions that are hidden to the user, such as storing a record in a database, transmitting information to another computer system to process an order, or emailing order information to a sales team.

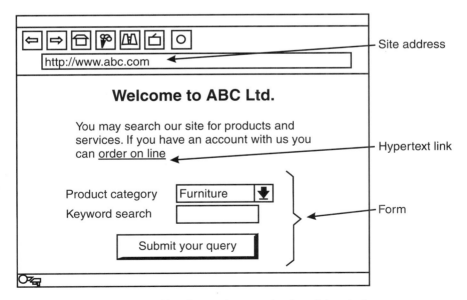

Figure 8.4 Web page viewed with an Internet browser showing a link and a form.

CGI scripts, while very useful, can potentially contain security holes if poorly written. For example, a poorly written script could be used by a hacker to obtain the password file from a web server. They should be only be written by somebody who fully understands the potential problems [2].

Secure Transactions on the Web

Most of the information transmitted between a web server and the browser is sent in plain text. There is, however, an optional encryption scheme, which is implemented in some servers and most browsers available today. This is desirable for the transmission of credit card numbers and other confidential information.

The user is normally given some indication that a secure transaction is taking place. In Netscape Navigator, for example, part of the screen changes color and a key symbol is displayed at the bottom of the screen. (The key symbol is always there but it is "broken" when not working in encrypted mode.) There are two levels of security available; 40-bit keys and 128-bit keys. This is further indicated by the key symbol having one or two prongs, respectively. The 40-bit key is considered insecure.

A new encryption key is automatically generated for each session with the secure server in a pseudo-random fashion. This key is transmitted to the user using public key encryption. This process is transparent to the user.

To protect against a hacker masquerading as a secure server and stealing your credit card number, the operator of a secure server can obtain an authentication certificate from a trusted third party (e.g., a *certification authority* such as VeriSign). This certificate is transmitted to the user at the start of a secure session. The browser can be configured to only accept certificates signed by specified certification authorities. Finally, the user can view the certificate with his or her browser, and if it contains the name of the company that he or she has connected to, he or she can be confident that the transaction will be safe [2].

Hosting Web Pages

It is very easy to advertise your company on the Internet. All that is needed is to pay an ISP to host your web pages on one of its servers. Creation of simple web pages is straightforward—most word processors have the capability to save files in web format. If however you need something more sophisticated, such as search facilities or online-ordering, you may need to get help. Many ISPs and other third parties offer such a service.

A collection of web pages owned by a particular company or organization is normally referred to as a *web site*. These pages are linked together using hypertext links. The starting point for a particular site is called a *home page*.

As already mentioned, you have the option of hosting your web pages on your own server or on an ISP's server. The relative advantages of each option are given next [3].

Hosting on an ISP's server:

- Will usually cost less, particularly for small businesses;
- Will usually have more bandwidth available;
- Will not need a permanent connection to the Internet to support the site;
- Will usually be professionally maintained with regular backups to tape and administrative staff available outside normal working hours;
- Will make it easier to protect a corporate network against unauthorized access (because access to the corporate network is not necessary).

Hosting on your own server:

- Allows you greater ease in creating links between your web server and information on other computers on your internal network;
- Allows you more control in protecting your web pages against hackers by putting your web server behind your firewall. The type of attack referred to here is where the hacker replaces your web pages with uncomplimentary or misleading pages.

Only organizations with considerable IT experience should consider hosting web pages from their own site. If you choose to use an ISP's server, you will need to check out exactly what you are getting:

- Who is sharing the server with you (if your web site shares a server with one or more very popular sites, they may slow down the web server)?
- How often is the server backed up, and are backups stored off site?
- Can you publish pages on the server without assistance from the ISP?

Search Engines

There are quite a number of powerful search engines on the Internet. These work by automatically gathering information from the web, indexing it, and storing this index on a powerful computer. Thus, when you use these search engines, you search this index, not the web itself. Search results can be returned within a few seconds. If you want to guarantee that your web site is included in these indexes you should register your home page with the search engine. This registration can normally be done for free over the web by filling out the appropriate form on the search engine's web site.

8.4.2 Email

Internet email is discussed in Section 10.3.1.

8.4.3 Telnet

Telnet is an application that allows you to connect to a computer (usually a UNIX host) in text mode. It can be used, for example, to access a text-based database that could not be accessed using a web browser. It is also a convenient way of allowing a network administrator to manually configure routers and servers. Because of telnet's text-only interface, most businesses prefer to make information available via a web browser.

8.4.4 News Groups

News groups are an important part of the Internet, though often neglected by newcomers. They are discussion areas where individuals pose questions or express opinions and others answer them. There are over 20,000 groups (in 1997), and a small proportion will be of interest to a business user. Communication and IT managers will find technical newsgroups covering most areas of interest. Depending on the nature of your business you may also find some news groups directly related to your core business.

News groups can be used to advertise your business, provided you go about it in a sensible way. It is generally acceptable, for example, to embed a link to your web page beside your signature in a message that contributes some useful information to the group. Blatant advertisements posted to a serious news group can result in the advertiser getting *flamed* (hate mail) or, worse still, mail bombed. Mail bombing means sending so many mail messages that the recipient is unable to use their mail account.

Many news groups are part of the *Usenet news* system. In Usenet, messages are replicated on a large number of *Usenet servers* around the world. Most ISPs, for example, maintain a Usenet server. This allows fast access to the messages but—due to space constraints—messages are only held on these servers for a couple of weeks. Archives exist for most news groups if you wish to look for older messages. Some Usenet groups are *moderated* (i.e., messages must pass through a person known as a *moderator*). The moderator may add pertinent comments and acts as a mechanism for keeping messages to the point.

Private news groups are also possible. These can be either open to all comers but not part of Usenet or they can have access restricted to those who pay a membership fee or to employees of a particular company.

8.4.5 Voice and Video

Voice and video can be transmitted across the Internet using a *real-time protocol* (RTP). While this is a very cheap way of communicating over long distances, there are two significant problems. First, there can be a significant delay (sometimes two or three seconds) in the communication path; second, the call-setup mechanisms are poor (at least at the time of writing calls are often established by prior arrangement or by using conventional telephony to inform the other party that you wish to converse over the Internet).

Voice is transmitted at about 8 Kbps, and speech quality deteriorates if there is insufficient bandwidth. Quality video requires more end-to-end bandwidth than the Internet can provide. A compromise is to reduce the picture size and reduce the number of frames per second.

8.4.6 Fax

Fax messages can be sent over the Internet using special fax servers. These servers can be connected to the PSTN as well as to the Internet, thus allowing faxes to be sent to conventional fax machines. An organization with Internet connections in different countries can save on international call charges for fax by installing an Internet fax server in those countries. Figure 8.5 shows the routing of a fax message from a PC on a LAN in one country to a fax machine connected to the PSTN in another.

As noted elsewhere in this book, email can, in many cases, replace fax. Even if the document does not exist in electronic format, it can be scanned. The scanned image can then be sent by email as an attached file.

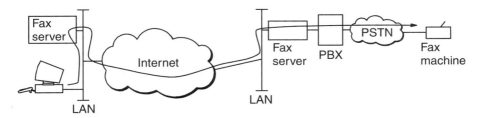

Figure 8.5 A fax message sent over the Internet.

8.4.7 Internet VPNs

The Internet can be used to interconnect LANs in different parts of the world. By using encryption, a *virtual private network* (VPN) can be created. The encryption is implemented in the firewalls, routers, or in a dedicated device. With some products, remote users, with the requisite software, can make a secure encrypted connection to the VPN. *Authentication* (I am who I say I am) is just as important as encryption when building Internet VPNs.

8.4.8 Intranets

An *intranet* is a corporate network based on Internet protocols and applications. Many companies have transformed their corporate networks into intranets because of the considerable price and usability advantages of Internet software.

Here are some examples of what can be done with a corporate intranet:

- Documentation can be published on the internal web.
- A search engine can be added to allow keyword searching of internal documents.
- Internal newsgroups can be established.
- Internet phone, fax, and video technology can be used to cut the communication bill.

Building an Intranet

An intranet consists first of an infrastructure of one or more interconnected LANs. Many organizations already have such an infrastructure in place and simply need to ensure that it is capable of handling the TCP/IP protocols. Chapter 12 describes the use of IP addressing and routing on such a network. The Internet can be used to provide the long-distance links for an intranet using encryption as described in Section 8.4.7.

Next you will need to establish at least one server to carry the web pages, search engines, name servers, and other server applications.

Finally, the desktop computers will need TCP/IP software with IP addresses and a web browser at a minimum. Depending on your applications, they may optionally require software for voice or video applications.

Extranets

The intranet concept can be further extended to the *extranet* concept, where trading partners agree on mutual access to servers across an IP network or the Internet to automate certain business transactions.

8.5 Addresses and Domain Names

Any medium to large business with a presence on the Internet will need a *domain name* such as ibm.com or telecom.ie. Such a name is somewhat analogous to a postal address. It allows data to find its way to your computers. Domain names need to be chosen carefully so that they are easily remembered (or better still can be guessed) and they create a favorable impression of your company. The latter point needs some explanation. A company called, say, ABC Ltd. in Ireland could opt for three different types of domain name:

- abc.ie—*ie* stands for Ireland;
- abc.com—*com* stands for commercial organization;
- abc.tinet.ie—*tinet.ie* is the domain name of their ISP.

Your chosen domain name will become part of your web address (e.g., www.abc.ie) and also part of your email address (e.g., sales@abc.ie).

The *abc.tinet.ie* option is the least advantageous. Besides being difficult to remember, it suggests that you are doing things on the cheap by piggybacking your domain name onto your ISP's domain name. It has the further disadvantage that if you change ISP, you will also have to change domain name. This could cause a disruption to your electronic communications similar to the disruption a change in your main telephone number would cause to your voice communications.

The *abc.ie* and *abc.com* names are the superior choices. They must, however, be registered for a fee with a registration authority. The .ie ending is the Irish *top-level domain* (TLD) and must be registered with the Irish registration authority. Each nation has its own national TLD. The .com ending is a *generic top-level domain* (gTLD) and should be used if your company engages in business around the world. Currently, it must be registered with the InterNIC in the United States. There are several gTLDs, most of which are self explanatory (e.g., .org [non-profit organization], .firm, .store) [4].

The registration process involves sending your chosen domain name along with details of your organization to the registration authority [5]. Names are allocated on a first come–first served basis. There is normally a dispute procedure that can be invoked for example when a trademark is usurped.

As a general rule it is best to hand over the registration process to your ISP for a small fee in addition to the registration authority fee. They must be involved in part of the process, and they will be experienced in the pitfalls. You should, however, ensure that you are named as the administrative contact [6].

Finally, domain names are not essential to have a presence on the web, and very small companies might prefer not to invest in the (modest) registration fee. In this case the web address of ABC might be *www.tinet.ie/companies/abc/*. This, of course, has the same disadvantages mentioned above for the *abc.tinet.ie* domain name.

At least one IP address will be required for a connection to the Internet. Your ISP should give you these addresses.

8.6 Security

There are a number of serious risks associated with connecting a corporate network to the Internet. The most important risk is from computer hackers who may have access to some or all of your computer systems via the Internet. The simplest protection against hacking is not to connect to the Internet. If you must connect, then you should implement a well-thought-out security policy including the use of a firewall at the point of access to the Internet.

8.6.1 Firewalls

A firewall is a computer or router connection between your network and the Internet. The firewall's function is to block unwanted traffic while allowing legitimate traffic through. It can block traffic on the basis of information in the packet headers, such as source or destination address, or the protocol in use (e.g., email, telnet, or web). So, for example, a firewall can be configured to block all access from outside to a particular host on the inside based on the IP address of that host.

The cheapest form of firewall is one implemented on the router that connects your network to the Internet. This type of firewall will perform the basic filtering required to keep most intruders out. It requires quite a bit of router configuration know-how, and expertise often has to be bought in.

A more comprehensive firewall is implemented in software on a dedicated computer attached to the same LAN as the router connected to the Internet. This type of firewall has the added advantage that it can detect suspicious activity (i.e., activity associated with known hacking techniques), raise alarms, and keep a log of particular events [7].

Care should be taken to install this second type of firewall on a computer that will be able to handle the traffic load. Remember that all traffic to and from the Internet must pass through this computer.

The firewall can also be used to enforce access rights for employees. Filters can be established in the firewall to specify what each computer on the internal network can and cannot access. This technique can be used, for example, to allow specific employees access to the web while restricting others to email only.

8.6.2 Access Control

No firewall system is perfect, and there is always a risk that an experienced hacker will penetrate your firewall. Further protection of your data can be achieved by logical access control to your computers. Passwords are a minimum, and their limitations should be understood. They can be hacked by eavesdropping, by obtaining password files, and through social engineering such as masquerading as a technical support person on the telephone and asking someone for their password. Other forms of social engineering include hackers taking temporary work with the company in question or with cleaning contractors.

Educating users can go some distance, but more sophisticated systems should be considered for highly sensitive data. One such technique involves the use of a device that looks like an electronic calculator, which generates an access code that changes every 60 seconds. The code is only valid during those 60 seconds. The user reads the code from an LCD display. The main point is that you have to have the physical device before you gain access [8]. If you lose it, you inform the network administrator, who cancels access from this device. Social engineering is thus rendered much more difficult. Authentication schemes using digital certificates (see Sections 8.4.1 and 8.6.3) can also be used.

8.6.3 Encryption

Any information sent over the Internet can potentially be intercepted along the way. This interception can be passive (read only), or if someone is really out to do damage they could possibly intercept and modify your information. Encryption is the best way to protect against this form of attack. There are three common encryption scenarios. First, email messages are encrypted in the sender's machine and decrypted at the recipient's machine. Second, transactions with secure servers are encrypted in the end machines. And third, when virtual private networks are created using the Internet, encryption is usually performed in the routers or in a dedicated encryption box connected alongside the routers.

There are two types of encryption in use today: symmetric key and public key. They have different characteristics and are often combined to get the best of both worlds.

Symmetric key is the simplest. A secret number called a secret key is used to encrypt the information using a scrambling procedure known as an encryption *algorithm*. The information can only be decrypted using the same key and algorithm. The algorithm is usually well known, and it is the secret key that provides the security. Both parties must know the key.

Public key encryption is more sophisticated. Each party has two keys, a public key and a private key. The public key can and should be freely advertised. The private key is never divulged. If Bob wants to send a message to Alice, he will encrypt it with Alice's public key. The only way to decrypt this message will be with Alice's private key.

It should be clear from the above that public key encryption simplifies the transfer of keys. It does have the significant drawback, however, that it is about 100 times slower to perform in software than symmetric key encryption. Most encryption schemes therefore combine the two methods as follows. Bob's software generates a random symmetric key and encrypts his message with it. It then encrypts this symmetric key using Alice's public key and sends it to Alice along with the encrypted message. Alice now has all she needs to decrypt the message. As added security, the symmetric key will be different for each message.

The strength of an encryption scheme depends on the length of key used. For symmetric key encryption, keys of 80 bits or more are recommended for valuable information. For public key encryption, the same protection requires keys of 768 bits or more [9]. As computer power increases, longer and longer keys will be needed.

Public key encryption can also be used to generate *digital signatures*. A digital signature is a data stamp appended to a message or file which quasi-uniquely links that file or document to the *private* key of the signer. Put another way, using a very powerful computer, it would take years to forge a digital signature. The digital signature can be verified by anybody using the *public* key of the signer. This verification not only verifies the origin of the message, it also proves the *integrity* of the message. Digital signatures are not yet legally binding in the United States but can still provide a high degree of confidence. Bilateral agreements can be used to provide legality in the absence of a generic law.

One problem with public key encryption is verification of the public keys. If Bob sends an encrypted and signed message to Alice looking for her credit card number, Alice must be sure that the public key that she assumes belongs to Bob actually does. One method of providing this guarantee is with a *digital certificate* provided by a trusted certification authority (as mentioned in Section 8.4.1). A digital certificate contains the name of the holder, his or her public key, an expiration date, and a digital signature signed by the certification authority. The public key of the certification authority can be trusted because it will be widely publicized by them. Hierarchies of trust can be established whereby, for example, a company certifies its employees and the company is in turn certified by the certification authority [10].

References

[1] Gareiss, Robin, "The Online Corporation: Choosing the Right Internet Service Provider," *Data Communications International,* November 21, 1995.

[2] Stein, Lincoln D., "The World Wide Web Security FAQ" Version 1.3.7, http://www-genome.wi.mit.edu/WWW/faqs/www-security-faq.html, May 7, 1997.

[3] Gareiss, Robin, "Web Hosting Services: No Mess, Less Stress," *Data Communications International,* November 21, 1996.

[4] International Ad Hoc Committee, "Recommendations for Administration and Management of gTLDs," http://www.iahc.org/draft-iahc-recommend-00.html, February 4, 1997.

[5] InterNIC, "New Domain Name Registration: Frequently Asked Questions," http://rs.internic.net/domain-info/registration-FAQ.html, May 1996.

[6] Wilson, Ralph F., "Does Your Business Need a Custom Domain Name?" *Web Marketing Today,* http://www.wilsonweb.com/articles/domain.htm.

[7] Newman, David, "Can Firewalls Take the Heat?" *Data Communications International,* November 21, 1995.

[8] Johnson, Johana Till, "Enterprise Security: Better Safe than Sorry," *Data Communications International,* March 1995.

[9] Schnier, B., *Applied Cryptography,* 2nd ed., New York: John Wiley and Sons, 1996, quoted in Chuck Shih, Mats Jansson, Rik Drummond, and Lincoln Yarbrough, "EDIINT Functional Specification—Requirements for Inter-operable Internet EDI," EDIINT Working Group, http://www.imc.org/ietf-ediint, March 1996.

[10] VeriSign, Inc., "Frequently Asked Questions about Digital IDs," http://www.verisign.com/repository/digidfaq.html, 1997

Further Reading

Cheswick, William R., and Steven M. Bellovin, *Firewalls and Internet Security—Repelling the Wily Hacker,* Reading, MA: Addison-Wesley, 1994.

Chapman, Brent D., and Elizabeth D. Zwicky, *Building Internet Firewalls,* Sebastopol, CA, O'Reilly and Associates, 1995.

9

Mobile Communication

9.1 Introduction

Mobile phones are no longer the status symbol nor the novelty they used to be. For many people today they are an essential part of work or leisure.

Anyone whose activities are not confined to a fixed location can benefit from a mobile phone.

- Tradesmen, for example, do not need to have someone employed to answer their office phone—in fact, they may not even need an office.
- Technical support staff can be contacted as soon as something goes wrong. This can mean a significant reduction in down time for all sorts of equipment, from computer networks to production lines.
- Sales personnel who are frequently on the move rely heavily on mobile phones because every lost call is a potential lost sale.
- Delivery staff can easily be alerted to any changes in a delivery schedule at short notice. This can save on delivery costs while improving customer satisfaction.

As mobile phones enter the consumer market, people are using them when they are out shopping, playing golf, or even while in school.

9.1.1 Mobile Phones and Cordless Phones

There is a certain amount of confusion as to the difference between a cordless phone and a mobile phone. Both use radio transmission to replace cables and hence look somewhat similar. A cordless phone has a base unit (a radio transceiver) attached to a normal fixed telephone line. It can be used around the house or apartment but cannot be used at any great distance from the base unit. This distance limitation ensures that one cordless phone does not interfere with another in a neighboring house or apartment. A mobile phone, on the other hand, can be used almost anywhere in the country in which it was purchased and often further afield.

9.1.2 Mobile Versus Wire Lines

Mobile phones give freedom from the fixed location of a wire-line telephone. As the cost of mobile communication falls, they will inevitably become the preferred option for many people. In the short term, mobile communication will augment rather than replace wire lines. Businesses, for example, will continue to use wire lines to connect PBXs and LANs to public networks. Wire lines are still a more suitable and cost-effective solution when it comes to high-speed data applications (30 Kbps and above), such as Internet access.

At the start of 1997, there were 135 million mobile phones in the world, compared to approximately 750 million wire-line phones [1]. The growth rate in mobile phones during 1996 was 59% per year compared to about 7% per year for wire-line phones [1].

9.1.3 History [2]

Mobile radio has been in use since the 1920s, when one-way radio was used to transmit messages to police cars. Two-way systems were introduced in the early 1930s. In 1946, AT&T made the first connection between a mobile service in St. Louis and the public telephone network. This system required an operator to connect the calls. The first direct-dial system was introduced in 1948. Up until the 1980s, most mobile radio systems were not connected to the public telephone network. They were used as a means of internal communication for police, army, utilities, government services, and the transport industry.

Introduction of Cellular

The biggest problem in the early days of mobile radio was congestion resulting from the limited number of radio channels. This problem was solved through the use of the *cellular* concept (see Section 9.2.2). On December 3, 1979, the world's first commercial cellular mobile radio service was opened in Japan. This was an analog system.

Analog Systems

In mobile telephony, the term *analog* refers only to the radio portion of the system. The rest of the system can be, and usually is, digital. During the 1980s many analog systems were developed. They are summarized in Table 9.1.

Introduction of Digital Systems

The biggest single development in mobile telephony in the recent past has been the introduction of digital systems. The primary objective was to increase the capacity of the systems. Table 9.2 summarizes some of the digital mobile radio systems in use today.

9.1.4 Developing Countries

Mobile telephony has a significant role to play in many developing countries where the fixed network is often poor. In some cases, mobile is the preferred solution for business because it provides the best or indeed the only telephone

Table 9.1
Analog Mobile Phone Systems

Name	Started	Main Countries of Operation	Approximate Number of Subscribers (1996)
Mobile Control Station (MCS)	1979	Japan	2 M
Nordic Mobile Telephone at 450 MHz (NMT450)	1981	Europe (especially Nordic countries)	2 M
Advanced Mobile Phone System (AMPS)	1983	North and South America, Canada, Australia	48 M
Narrowband AMPS (NAMPS)	1992	South America and Asia	2 M
Total Access Communication System (TACS)	1985	Europe (especially U.K. and Italy)	14 M
Nordic Mobile Telephone at 900 MHz (NMT900)	1986	Europe	3 M
Japanese Total Access Communication System (JTACS)	1989	Japan	2 M
Narrowband Total Access Communication System (NTACS)	1991	Japan	2 M

From: [3,4].

Table 9.2
Digital Mobile Phone Systems

Name	Where Used	Started	Further Details
Global System for Mobile Communication (GSM)	Europe, Australia, others	1992	140 million users in mid-1997
Digital AMPS (D-AMPS)	U.S., others	1992	4 million users in 1996
Digital Communication System at 1800 MHz (DCS1800)	Europe	1993	Also known as PCN
Personal Communication System at 1900 MHz (PCS1900)	U.S.	1995	Based on GSM
Personal Digital Cellular (PDC)	Japan	1993	Was called JDC, 10 million users in late 1996
Personal Handyphone System (PHS)	Japan	1995	Telepoint, 5 million users in early 1997
Code-division multiple access (CDMA)	U.S., others	1995	First used in Hong Kong

From: [3–6].

service. For example, when granting a cellular license to E-Plus in Germany in 1993, the licensing authority stipulated that priority be given to providing service in the former East Germany where the land-line infrastructure was very poor [5].

Many severely underdeveloped countries, however, have very poor or in some cases nonexistent cellular services. India and many central African countries, for example, had no cellular services up to 1995.

9.2 Principles

9.2.1 Radio Communication Jargon

What Are Radio Waves?

Mobile communication is based on radio. Radio waves are *electromagnetic waves*. Ordinary light, radar, and x-rays are also electromagnetic waves. These different types of electromagnetic waves are distinguished by their frequency (i.e., the rate at which the electrical and magnetic fields vary with time). For example, radio waves have lower frequencies than light waves, which in turn have lower frequencies than x-rays (see Figure 9.1).

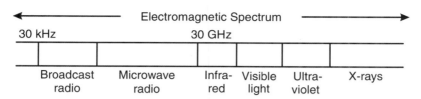

Figure 9.1 Radio in relation to other electromagnetic waves.

Frequency

Frequency is measured in hertz (Hz) where 1 Hz is equal to one cycle per second. For example, you could say that a pendulum swings with a frequency of about 1 Hz. Radio waves, on the other hand, have very high frequencies—in the range of 30 kHz to 30 GHz.

At present, mobile phones use frequencies close to 450, 800, 900, 1,500, 1,800, and 1,900 MHz. Figure 9.2 shows that all of these frequencies lie within the *ultrahigh frequency* (UHF) band; however, they are often referred to as microwaves.

Spectrum

The word *spectrum* is used to indicate a broad range of frequencies. We can use this term to refer to the entire electromagnetic spectrum or a particular portion, such as the radio spectrum.

Spectrum Is Limited

Other services, such as satellite communication, point-to-point radio links, and TV broadcasting must all get their share of radio spectrum along with mobile

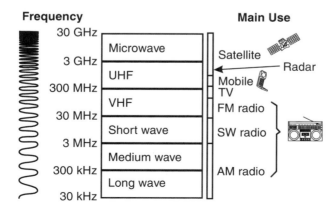

Figure 9.2 Radio bands.

radio (see Figure 9.2). This means that maximum use must be made of the available spectrum if mobile phone systems are to be able to support millions of users.

Bandwidth and Channels

Transmission of voice or data over radio waves requires a radio channel. In broadcast radio, for example, different radio stations each have their own channel positioned at intervals along the dial of the receiver. The difference between the highest and lowest frequency in a radio channel is called the *bandwidth* of the channel. This is a measure of how much spectrum the channel uses. For example, the AMPS mobile phone system uses 30 kHz per radio channel, and each radio channel carries one speech channel. The D-AMPS system is more efficient—30-kHz radio channels are used, but each radio channel carries three speech channels.

Figure 9.3 shows how the 800-MHz band is divided into channels for the AMPS system. There are a total of 1,664 radio channels, which translates into 832 duplex channels (because speech requires a separate transmit and receive channel).

Allocation of Radio Frequencies

Radio frequency usage is governed by national and international organizations. In the United States, it is controlled by the *Federal Communications Commission* (FCC). Internationally, it is governed by the ITU. Frequency allocations are not the same in every country, although efforts are made to harmonize them.

The use of different frequencies in different countries makes it difficult to introduce a single mobile phone system that can be used in every country. For example, in the United States, most mobile networks operate in the 800- and 1,900-MHz bands, while in Europe, the 900- and 1,800-MHz bands are allocated to mobile telephony. This is one reason why American and European mobile phone systems are incompatible. Another reason is that the systems were

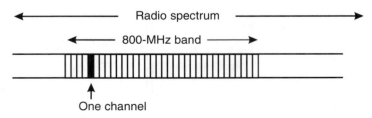

Figure 9.3 Radio channels occupy a small range of frequencies within the frequency band.

designed differently with different channel bandwidths and different signals sent between the network and the phones.

Line of Sight

Radio waves travel in a straight line. This gives rise to the term *line of sight,* where the communication can only occur if the transmitter can be seen from the position of the receiver. The radio waves used in mobile phones reflect off solid objects such as buildings. These reflections have both positive and negative effects. On the plus side, they allow signals to be reflected into areas that are not within line of sight of the main antenna. On the negative side, they lead to problems if the obstructions are too great, such as behind hills, inside tunnels, and, to a lesser extent, inside buildings. Reflections also lead to an effect known as *multipath interference,* where reflected signals interfere with the main signal and degrade the signal quality in certain locations. In digital systems the signal quality can vary considerably over a distance of less than a meter.

9.2.2 The Cellular Concept [2]

Modern mobile radio systems are *cellular.* This means that the coverage area is divided up into small areas called cells. Each cell has its own antenna for transmitting to and receiving from the mobile phones. This allows the network operator to reuse radio channels in cells that are sufficiently far apart. This greatly increases the network capacity (Figure 9.4).

For a given technology, the normal way to increase the capacity of the network is to increase the number of cells while reducing the size of the cells

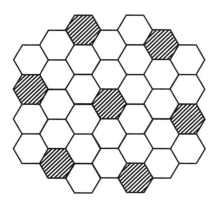

Figure 9.4 Splitting the coverage area into cells allows radio channels to be reused in cells that are sufficiently far apart from each other.

(Figure 9.5). Cells are therefore much smaller in towns and cities than in rural areas to accommodate the higher density of users.

Capacity can also be increased by allocating more spectrum to the service or, in digital systems, by using lower bit-rate voice coders.

The Mobile Network

Each cell has an associated base station and antenna. The base station consists of the radio equipment, which is linked back to a *mobile switching center* (or *mobile telephone switching office* [MTSO] in the U.S.) via digital transmission links similar to those used in the PSTN. Optic-fiber cables, copper cables, or microwave links can all be used to link the base station back to the switching center (Figure 9.6). The mobile switching center is a telephone exchange that has been designed to handle all of the functions required in a mobile telephone network. It is connected to the PSTN/ISDN and to other mobile switching centers.

Handoff

If a mobile user in a car moves from one cell into another during a call, the network must be able to detect this fact and transfer the call to the new cell. The network detects the movement by measuring the strength and/or quality of the radio signal coming from the mobile. Alternatively, the mobile phone can make the measurements and inform the network when the signal is getting weak. The network reacts by looking for a suitable cell (one with a strong signal)

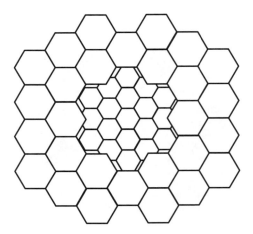

Figure 9.5 Increasing the cell density allows more mobile phones to be used in a particular area.

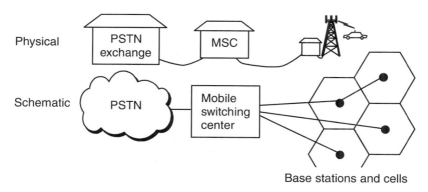

Figure 9.6 Base stations linked back to the mobile switching center.

and transfers the call to a free channel in that cell. This procedure is known as a handoff. In most mobile radio systems, handoffs cause a minor disruption in service, which is quite tolerable for voice calls but can be a problem for data transmission.

Speed of the Mobile Phone Implies a Lower Limit to Cell Size

Mobile phone networks are designed to support mobile phones moving at high speed. For example, GSM specifies a maximum mobile speed of 250 kph (170 mph) [7]. (One assumes that they had high-speed trains in mind and not sports cars.) The biggest problem posed by a high-speed mobile is handoff. If the cell selected for the handoff is very small, there is a chance that the mobile will have passed through the cell before the handoff is accomplished. Special arrangements, described next, must be made if very small cells (less than 300m) are required in a particular area.

Microcells

Cells with a diameter less than about 300m are called microcells. They are usually implemented by locating the base station antennas below the rooftop level and using very low transmit power. The area covered by a number of microcells should also be covered by an umbrella cell or macrocell (Figure 9.7).

Arrangements are made to connect slow-moving mobiles to the network via the microcell base station and to reserve the larger umbrella cell for high-speed mobiles. The network can detect high-speed mobiles by measuring the rate of change of signal strength over time or by measuring a phenomenon known as Doppler shift.

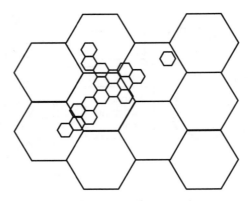

Figure 9.7 Areas of high mobile density served by microcells and by macrocells.

9.2.3 Analog Versus Digital

The move towards digital systems is motivated primarily by the requirement for extra capacity. There are other important objectives:

- To improve privacy;
- To improve security against fraud;
- To increase compatibility with the fixed network, and in particular ISDN;
- To improve speech quality in areas of poor reception.

Phasing Out Analog Systems

Many analog systems are likely to be phased out over time, particularly in areas of high population densities. Some countries have already announced closure of their analog networks. The AMPS network in Australia, which had 2 million connections in mid 1995, will be discontinued after the year 2000 [3]. An NMT450 network (30,000 connections) in Belgium will be discontinued after 2000 [3]. In Spain, the NMT450 network will close in January 1998, and the Spanish TACS network (1.4 million connections in 1996) is scheduled to close in 2006 [3]. Some handset manufacturers have already ceased production of analog handsets.

The phasing out of an analog network will in general require a number change and a handset change for the customers. This is one more incentive for new customers to opt for digital service. The phasing out of AMPS in the United States (not an imminent proposition) is likely to be a smooth transition, as it can share its numbering scheme with a D-AMPS system. It will, however, mean that people with older handsets will have to replace them.

9.2.4 Speech Quality

Compared to land lines, all mobile phone systems suffer from some degradation in speech quality. A digital mobile network does not necessarily imply superior voice quality and can certainly not be compared to the digital quality available from a compact disc. This is because the bit rate used to code speech in digital mobile radio is less than 1/50th of that used to code the sound on a CD. Analog systems in areas of good reception can give superior quality to digital systems. Digital systems, on the other hand, maintain their speech quality to a greater degree than analog in areas of poor reception or high interference. The voice coders for most digital mobile radio systems are a compromise between quality and bit rate. Many users of digital mobile phones claim that they were accustomed to superior quality when using analog services.

Deficiencies in voice quality have the following causes:

- Poor signal quality due to location (i.e., far from a base station or inside a large building).

- Poor network: Mobile networks require careful planning so that the effects of interference from other mobile phones and base stations are kept to a minimum. It was alleged at one stage, for example, that the voice quality on both GSM networks in the United Kingdom did not match that of the GSM networks in the rest of Europe [8].

- Poor mobile equipment: Some phones perform much better than others both in terms of clarity and loudness. Some hands-free systems cause echo on digital systems because the sound from the speakers is fed back into the microphone. This effect can be minimized through the use of a good directional microphone.

- Digital voice coders: The voice coders in digital cellular employ low bit-rate coding. The bit rate is kept low to maximize network capacity. While the speech is quite intelligible, many people refer to it as artificial, metallic, or like a robot. The industry has taken note, and enhanced voice coders have been developed for GSM and D-AMPS. At the time of writing, however, these are not widely used.

Half-Rate Voice Coding

Most digital systems are designed so that half-rate voice coders can be introduced at a later date. As the name suggests, half-rate voice coders only require half the digital bit rate of a full-rate coder to transmit speech. Half-rate coding has the net effect of keeping speech quality at the same acceptable level while doubling the network capacity.

There are indications that half-rate coding will be implemented in very few mobile networks on account of the increasing demands for quality from the users. Any future changes to voice coders are likely to be in the direction of increased quality rather than reduced bit rates.

9.2.5 Charging

There are two very different ways of charging for calls made *to* a mobile phone. In one case, the caller pays the full amount for the call. The other method, which is the norm in the United States, is for the receiver of the call to pay for cellular *air time* while the caller only pays the equivalent land-line charge. This option is attractive for salespeople and others who depend on incoming calls to get business. Both methods have their advantages, and some cellular operators in the United States are offering "calling party pays" accounts to their mobile customers. Conversely, it is possible to get a freephone (800 and 888 services in the U.S.) number directed to your mobile phone. This might be useful way of attracting more business via a mobile phone, particularly in Europe where the caller has to pay a lot for a "local" call to a mobile phone.

Per-Second Billing

Another point to watch is whether a particular provider charges by the minute or by the second (or something in between—6 seconds is common in the U.S.). Per-minute charging can be very costly if a lot of short calls are made.

Tariff Plans

In competitive markets, mobile operators offer a range of tariff plans. For example, a plan designed to suit those who only use their phone occasionally would have a low fixed monthly rental combined with high call charges, while a plan to suit a high-volume caller would have a high rental but a low call charge. Tariff plans can also be distinguished by the amount of free charge units included in the monthly rental. It can be very confusing for a newcomer, and many operators allow their customers to change plans at no extra cost.

Itemized billing is a must when it comes to checking if you have chosen the correct tariff plan.

Calling Freephone Numbers

Not all network operators allow freephone numbers to be called for free. For example, you may have to pay for the cellular air time.

Handset Subsidies

Many handsets are subsidized by the network operators to reduce the acquisition costs for new customers. In the United Kingdom, for example, this works through agents who sell the phone along with a connection to a particular network. To protect their investment, the agents can "lock" the phone to that network, and if the customer wants to move to another agent or network they must pay a fee to have the phone unlocked. In the United States, service contracts of two to three years are required to qualify for a subsidized handset.

9.2.6 Coverage

A big issue, in choosing a mobile network, is coverage. Very few parts of the world can claim a mobile network with 100% population coverage. Full *landmass* coverage is even harder to find. The main obstacles to achieving full coverage in any country are:

- Areas of low population density;
- Areas with difficult terrain in terms of line-of-sight coverage;
- Difficulties in gaining planning permission for cell sites.

In many cases, 90% coverage can be achieved less than five years after the granting of a license. In Germany, E-Plus achieved 90% coverage of former East Germany (109,000 sq. km—about the size of Pennsylvania) by the end of 1995, just under three years after they were granted their license. In the same time period, they had achieved 75% coverage of the whole of Germany (357,000 sq. km) [9].

Rapid network development is normally spurred on by competition from other operators and by the attraction of a large market. Such is not generally the case in the poorer parts of the world.

9.2.7 Security [5]

There are three main security issues associated with mobile phones:

- Fraudulent use of somebody else's account (toll fraud);
- Theft and subsequent resale of handsets;
- Eavesdropping.

Security in Analog Systems

The original analog systems were, and to some extent still are, chronically poor at protecting against these abuses. Newer systems, and in particular digital systems, have benefited from hindsight and include more sophisticated security measures.

It is unfortunately relatively easy to obtain the *electronic serial number* (ESN) of an analog phone through the use of a radio scanner. It is similarly easy to reprogram most phones with this ESN. The process of reprogramming a phone is called *rechipping*. By this means, it is possible either make calls on someone else's account (this is referred to as cloning their phone) or to rechip a stolen phone so that it is no longer detectable as stolen.

A number of initiatives have been taken to improve the security of analog systems. In April 1995, Vodafone in the United Kingdom implemented TACS authentication. NYNEX in the United States has introduced PIN numbers, which must be keyed every time a call is made. The FCC has ruled that the ESN numbers in new phones must be tamperproof. This FCC ruling will make new phones very unattractive to steal but will not prevent cloning of their ESNs in old phones.

Finally, it is very easy to eavesdrop on analog mobile calls using an adapted mobile phone.

Security in Digital Systems

The GSM, DCS1800, and PCS1900 systems use a sophisticated authentication scheme in a smart card to protect against toll fraud. The handset manufacturers are required to make the *international mobile equipment identifier* (IMEI)—the equivalent of the ESN—physically secure and uneconomic to change, thus making handset theft uneconomic. The radio transmission is encrypted to prevent eavesdropping of both voice and data calls.

Networks that allow roaming need to be able to check whether a roaming handset is blacklisted. This is solved in the case of GSM and DCS1800 through a central database located in Dublin, Ireland, which acts as a collection and distribution point for blacklisted numbers of all participating network operators.

9.3 Network Services

9.3.1 Roaming

Roaming refers to the possibility of using a mobile phone while in the coverage area of a different network (e.g., when traveling abroad). The ability to roam from one network to another requires that the same or compatible technologies are used in the two networks *and* that a roaming agreement exists between the two network operators. For example, it is possible to roam between many of the AMPS-based mobile networks in the United States, but it is not possible for a U.S. subscriber to roam in Europe because none of the European mobile phone systems are compatible with AMPS. Up until 1996, only networks based on AMPS, NMT, and GSM have had a significant number of roaming agreements and only with networks of the same technology. Table 9.3 gives a broad picture of international roaming agreements.

Through the use of *dual-mode phones* (phones capable of working on two different systems) or *dual-band phones* (phones capable of working in two frequency bands), roaming is sometimes made possible between different technologies, as indicated in Table 9.4. A method of roaming between different GSM-based technologies is called *SIM roaming*. This is achieved by transferring a *subscriber identification module* (SIM) card (also known as a smart card) between phones capable of working on the different technologies.

Roaming Charges

There is often a premium charged for calls made and received while roaming. In addition, the receiver normally pays for any international portion of the call

Table 9.3
Main Roaming Agreements Throughout the World (1997)

Area	Main Mobile Technologies	Main Roaming Agreements
U.S.	AMPS	Within U.S. and to Mexico and Canada, PCS 1900 to GSM roaming emerging
Canada	AMPS	Within Canada and to U.S.
Europe	GSM, DCSNMT, TACS	GSM roaming within Europe and to other GSM networks; DCS roaming within Europe; limited NMT roaming
Japan	PDC, MCS, JTACS	National roaming
Australia	GSM, AMPS	GSM roaming, mainly with Europe
Latin America	AMPS	Nationwide roaming in most countries
China	GSM, TACS	Mainly within China
Russia	GSM, NMT-450, AMPS	Within Russia, plus GSM roaming mainly with Europe
Africa	GSM, TACS	Very little roaming

From: [6].

Table 9.4
Compatible Mobile Systems (Requires Suitable Phones and Roaming Agreements)

System	Roams With
D-AMPS 800	AMPS (dual-mode phone)
D-AMPS 1900	AMPS, D-AMPS 800 (dual-mode, dual-band phones)
CDMA 800	AMPS (dual-mode phone)
CDMA 1900	AMPS (future possibility, dual-mode phone)
PCS1900	GSM and or DCS1800 (SIM roaming, dual-mode phones expected in 1998)
GSM	DCS1800 (SIM roaming or dual-mode phone)

charge that the caller might be unaware of. For example, a U.K. subscriber roaming in Australia receives a call from a friend in the United Kingdom. The caller pays the national call rates but the receiver must pay for the international portion of the call plus any premium for the roaming facility.

The call charges while roaming can be quite unexpected at times. Let's take our U.K. subscriber roaming in Australia again, but this time she receives a call from somebody residing in Australia. The caller dials a U.K. number and pays an international call to the United Kingdom. Meanwhile, the receiver pays for an international call from the United Kingdom to Australia. This is because the call will actually be routed via the United Kingdom.

This situation can be overcome through the use of common channel signaling in the international telephone network. In this case, a signaling message would be sent to the United Kingdom. inquiring the location of the mobile phone. A signaling message would be returned to inform the Australian telephone exchange that the phone is roaming in the Australian network and the call would be connected internally in the Australian network. This arrangement, when implemented, will allow realistic call charges to be levied for this call.

National Roaming

National roaming means roaming between two or more similar networks in the same state or country. This can be useful if the two network operators concerned have different coverage areas. One network effectively fills out the other's black spots. Again, a premium may be charged for calls made while roaming in this manner. In Canada, for example, there are more than 10 different AMPS networks, but nationwide roaming is possible between these networks.

Seldom do you come across two operators having the same nominal coverage area allowing roaming between their rival networks.

9.3.2 Data and Fax

A mobile phone can be connected to a laptop computer to give extra mobile office facilities, such as fax, email, remote login, and an Internet connection. *Personal digital assistants* (PDAs) and small palmtop computers are a cheaper/lighter alternative. The equipment required to connect the phone to the computer varies, and some mobile phones do not support this type of connection at all. In addition, some digital cellular networks do not support data communication. For example, the standard for data on D-AMPS has only recently been ratified and is not yet implemented in many D-AMPS networks. Fortunately, because most D-AMPS phones are dual mode, data can be sent by switching to analog mode. A growing number of products integrate the phone with a PDA.

Analog Cellular Modems

An analog phone requires a cellular modem and a special interface cable. Cellular modems are available on PCMCIA cards. These are the credit card–sized cards that slot into laptop computers. A cellular modem is similar to a land-line modem and does not cost a whole lot more. It has special protocols, such as ETC or MNP10, built in to combat the momentary disruptions in service associated with cellular networks. Most cellular modems are dual standard (i.e., they can be used on land lines or on mobile phones).

To get the most out of these cellular modems, the same protocol must be used at both ends of the connection. For example, a business that wanted its mobile employees to be able to dial into their computer system should install cellular modems in headquarters. An Internet provider might have a special pool of cellular modems for mobile callers. Many cellular providers have installed cellular modem pools in their network. If you prefix your data calls with a special code, the call will be routed via one of these modems and the correct protocols will be used over the most demanding part of the connection (i.e., the radio connection between your phone and the base station). With such a service you do not need to worry about the type of modem at the far end of the connection.

It is usually better to force fax connections to use 4.8 Kbps or less. This is going to cost more, but it increases your chances of success. Receiving a fax via a mobile phone can be problematic. It is quite common for calls to fail during the initial handshaking. Sometimes the connection can be made to work by forcing a lower speed connection.

Digital Data Cards

Digital phones require a special fax/data card. In 1997, these are somewhat more expensive than analog cellular modems.

Bit Rate

Data communication over a mobile phone is not nearly as fast as the 33.6 Kbps possible over a land line. GSM-based systems (GSM, DCS1800, PCS1900) allow communication at up to 9.6 Kbps. They will also hold a connection while traveling at high speed. Analog systems such as TACS and AMPS will connect at a speed ranging from 1.2 to 14.4 Kbps, with 4.8 Kbps being about average in the United States [10]. Connections at 14.4 Kbps require ideal conditions (i.e., a good signal with very little interference). Analog data works well if the phone is stationary but is very troublesome when traveling in a car or train.

The new digital PCS systems in the United States will offer data services at 9.6 Kbps. Some CDMA systems will be capable of 14.4 Kbps. Higher bit rates may become available on these systems in the future by combining or aggregating two or more channels for one call [11].

All mobile data connections will be subject to high *error rates* caused by interference in the radio connection and other effects. Thus, even if a connection is established at 9.6 Kbps, the effective throughput can be much lower than this—the reduction being caused by the need to retransmit blocks of data found to be corrupted. In digital systems, there are two data transmission modes: *transparent* and *nontransparent*. Nontransparent mode virtually eliminates errors on the radio link by incorporating a special error-correction protocol between the mobile and the network. This protocol is optimized for the high error rates often found on digital mobile radio connections. On poor connections, nontransparent mode is superior to any other error-correction mechanism.

9.3.3 Short Message Service

Most digital mobile technologies include a facility for sending and receiving short messages, typically 160 characters in length. Unlike many existing paging services, the network can acknowledge receipt of messages. Messages are sent via a message center owned by the network provider. If the message is undeliverable, either because the recipient is out of coverage or has switched off the phone, the network can store the message until it becomes possible to transmit. Messages can be sent from a mobile phone via the keypad or an attached computer. It is also possible to send the message to the message center via the PSTN, ISDN, packet-switched network, or the Internet.

9.3.4 Video

Video transmission requires a lot more bandwidth than currently available on a standard mobile phone connection. Techniques are being developed to combine (aggregate) the bandwidth from a number of channels to give the required bandwidth for video. Using today's technology, however, this would involve combining four to eight channels to give a fairly poor quality image. It is doubtful if many users would pay for such a service if it were charged at four to eight times the cost of a normal cellular call.

9.3.5 Vehicle Tracking

Tracking the location of delivery vehicles or customer service staff can help to cut down transport costs and reduce delays by allocating jobs to the person who is best positioned to do it. Location updating is already possible through the use of satellite-based GPS equipment. The GPS information can be transferred to a computer and transmitted as a short message back to a control center via a mobile phone or mobile data terminal at regular intervals. Some satellite-based data messaging systems (e.g., EutelTracs) can have a GPS receiver integrated into the mobile messaging terminal.

9.4 Handset Technologies

9.4.1 Handset Features

The following features have already been implemented in some models of mobile phones and are features worth considering when deciding on a model of mobile phone:

- Built-in mechanism for tracking call charges;
- A mechanism to prevent calls from being initiated while the phone is in your pocket or bag—either a keylock (i.e., a key sequence that locks the keys or a "flip" mouthpiece that covers the keys);
- Built-in voice mail—avoids the need to call the network voice-mail system to retrieve messages;
- Single key access to voice mail;
- Auto answer after one ring is a useful feature if you intend to use the phone in hands-free mode in a car;
- PIN security on the phone deters thieves and helps avoid fraud if the phone is stolen;

- Built-in phone book with names and numbers;
- A flexible call-barring mechanism (e.g., bar all numbers except the first four numbers in the built-in phone book)—editing of the phone book should also be barred in this instance;
- Ease of use—large screen, easy menus;
- Dual band or dual mode (see Section 9.3.1).

Handset facilities that require network compatibility include:

- Short message capability;
- Data and fax capability;
- Enhanced speech.

9.4.2 Hands-Free Operation

From a safety perspective, mobile phones should be used *hands free* when driving. Kits that will convert a handheld phone into an in-car, hands-free phone are available. These kits normally include a directional microphone, a speaker, and an external antenna.

9.4.3 Power Class

All mobile phones have a maximum power output. The higher the output, the farther a mobile phone can be from a base station and still maintain a connection. It is not surprising that the coverage of a cellular network is better for phones with a higher maximum power. In general, handheld phones have the lowest power; next comes the portable phone with its bulky battery and transmitter. The highest powered units are those installed permanently in a vehicle.

All mobile networks have an upper limit on the permissible maximum power output from the phones. It is here that we find the biggest difference between the original cellular networks (AMPS, D-AMPS, TACS, NMT, GSM) and the latest PCNs or PCSs. In the latter case, the maximum power output from the phones is much more restricted. In these networks, handheld phones can have as good a coverage as vehicle-mounted phones.

9.4.4 Batteries [12]

The talk time and standby time of a mobile phone are related to the battery capacity. There are three battery technologies popular today—nickel cadmium

(NiCd or "Ni-Cad"), nickel metal hydride (NiMH) and lithium ion (Li Ion). They are listed here in order of increasing capacity-to-weight ratio. For a given capacity, a lithium ion battery weighs about half that of an NiCd but is more expensive.

NiCd batteries suffer from a problem referred to as memory effect. It comes about when the battery is partially discharged to the same level before recharge on a regular basis. Eventually the battery will appear to have a reduced capacity. It is as if the battery remembered how much it was discharged before recharge and will only discharge to that level the next time. To avoid this situation, a NiCd battery should be discharged fully before recharging. Memory effect can be reversed by giving the battery a few cycles of full discharge followed by full recharge. Memory effect is not present in NiMH or Li Ion batteries.

Cheap battery chargers (typically those that come with the phone) use a simple trickle-charge mechanism, which typically takes 8 to 12 hours to charge a battery. Rapid chargers that can do the job in as little as an hour are available. They cost more because they have a lot more electronics but are far more convenient. Very-high-speed charging (one hour and less) is usually less effective than a slower charge (2 to 4 hours) because it upsets the battery's chemistry somewhat. Be sure to purchase a reputable charger designed for your battery model—if a battery becomes overcharged it will be damaged.

9.5 Networks Technologies

The following is a description of the various technologies used in mobile phone networks. There are three main sources of standards for mobile telephony corresponding to the three largest markets: Europe, the United States, and Japan.

9.5.1 Europe

Nordic Mobile Telephone System

This system gets its name from the four Nordic countries of Denmark, Finland, Norway, and Sweden, which jointly developed the specification. The first NMT system was launched in 1981. It is now in use in about 40 countries, mainly in Europe, North Africa, and Russia [3]. There are two variants, NMT450 and NMT900, operating in the 450-MHz band and the 900-MHz band, respectively. Dual-band phones are available. NMT450 is still popular today in regions of sparse population because it allows very large areas to be covered by a single base station.

Total Access Communications System

TACS is an analog cellular system used mainly in the United Kingdom and Italy. It operates in the 900-MHz band. An extension to the original channel allocation in the 900-MHz band gave rise to the E-TACS (extended TACS) name. There were about 14 million users worldwide in 1996.

Global System for Mobile Communications

GSM was initially designed as a pan-European system. The initial feature set included international roaming, short message services, and data services. GSM is not, however, confined to Europe, and by mid-1997 it was in use in about 100 different countries around the world.

GSM operates in the 900-MHz band and competes for bandwidth with NMT900 or TACS systems (Figure 9.8). This makes it more difficult to plan and build a GSM network in those countries that already have TACS or NMT900, particularly when the limited bandwidth for GSM has to be shared between two operators.

The SIM Card

GSM uses a smart card called a SIM to "personalize" any GSM phone. It stores many details, including your mobile telephone number, a personal phone book, and your authentication key. It allows you to use *any* phone to make and receive calls. Outgoing calls will be billed to your own account. Incoming calls are directed to the phone that has your smart card inserted.

Whenever you use a GSM phone, the network checks to see if you are using a valid SIM. An authentication key is stored on the SIM. A challenge-response procedure is used to authenticate the SIM. The challenge-response mechanism never transmits the authentication key over the radio waves, making it very difficult to fraudulently use somebody else's account.

Data

Data and fax transmission is possible between GSM and the ISDN/PSTN networks at rates from 300 bps to 9.6 Kbps. A higher bit rate of 12 Kbps is possible

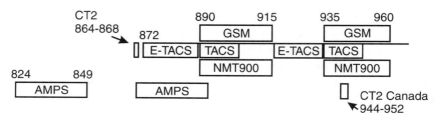

Figure 9.8 Spectrum allocation in the 800- and 900-MHz bands.

between one GSM handset and another [7]. The ability to use data and fax is dependent on both the network provider and the phone. The network operator will determine what bit rates if any are actually supported.

A *short message service* (SMS) allows a message of up to 160 characters to be sent to a GSM phone, even when there is already a call in progress.

DCS1800

DCS1800 (also called GSM1800) is a variant of the GSM standard. The following differences between the systems are present:

- DCS operates at a higher radio frequency (1,800 MHz);
- DCS has three times as much bandwidth as GSM allocated in Europe;
- DCS networks will be optimized for low-power handheld phones (max power = 1W).

Although the DCS1800 standard was available very soon after GSM, the allocation of licenses has been much slower. Given the larger bandwidth allocation for DCS1800, it could become more widespread than GSM in the more densely populated regions of Europe in the long term.

Dual-application SIMs and dual-mode phones are available to provide roaming between GSM and DCS1800 networks.

9.5.2 United States

Advanced Mobile Phone System

Cellular radio was introduced in the United States in 1983 using a single technology known as the *advanced mobile phone system* (AMPS) throughout North America. This facilitated the introduction of roaming between the various networks. AMPS is also used in Canada, Latin America, Australia, and parts of Asia. It is the most widely used mobile technology in the world (mid-1997).

Digital AMPS

To increase the capacity of AMPS, the *digital AMPS* (D-AMPS) standard was developed. Using time-division multiplexing, it puts three voice calls in the same radio channel that AMPS uses for a single call. D-AMPS can be integrated into existing base stations without the need for new antennas. Dual-mode phones are also available (i.e., AMPS/D-AMPS phones). This allows a smooth transition between the technologies [12].

There are two D-AMPS standards: IS-54 (1991) and IS-136 (1994). The latest standard supports authentication, calling line identification, voice encryption, data, text messaging, and fax services [5].

The voice coder specified for D-AMPS operates at 8 Kbps using technology that is somewhat more advanced than that used in GSM. The voice quality is roughly equivalent to GSM (i.e., somewhat poorer than that experienced on a good land-line phone). Enhanced voice coders may be introduced in the future.

Personal Communication Service

The latest breed of mobile services in the United States is called *personal communication services* (PCS). (PCS also refers to paging services.)

The FCC began auctioning spectrum in the 1,900-MHz band for cellular PCS in 1995. In all, 140 MHz of bandwidth will be allocated to PCS mobile radio, and 20 MHz of this bandwidth will be unlicensed. This compares with 150 MHz allocated to DCS1800 in Europe.

Unlike the situation in Europe, no single technology is specified. Instead the license winners are planning to use a number of different technologies. Three technologies have already been chosen by the license holders:

- D-AMPS 1900;
- PCS1900;
- Code-division multiple access (CDMA).

D-AMPS 1900.   This uses the latest version of the D-AMPS 800 standard, IS-136. The only difference is the radio frequencies used. Dual-band systems and dual-band phones are both available. Dual-band in this case means that D-AMPS 1900 systems can be built using the same infrastructure as D-AMPS 800 and that the phones will be able to work in both systems. This gives the PCS subscriber the advantage of the wide coverage already provided by the 800-MHz services from the day the PCS service is launched, provided the PCS operator has roaming agreements with the 800-MHz services.

PCS1900.   PCS1900 (also called GSM1900) is based on the GSM and DCS1800 standards developed in Europe. It is a commercially proven system and has been chosen by cellular operators that believe that time to market is the most important factor in the choice of a technology to use for their PCS licenses.

CDMA.   CDMA is a spread-spectrum technology that has been used in military applications for many years but is relatively new to cellular radio. CDMA is claimed to be a superior technology with a number of benefits over the other two systems:

- Spectrum efficiency—CDMA is claimed to support at least three times the number of customers for a given cell site distribution and spectrum allocation.
- Soft hand-off—CDMA systems allow a "make before break" handoff with no momentary disruption in the signal.
- A frequency reuse factor of one—the same frequencies can be used in adjacent cells. This eliminates the need for frequency planning and facilitates the soft handoff technique.

The North American CDMA standard is IS-95 (1993). The world's first commercial CDMA system was deployed in Hong Kong at the end of September 1995.

9.5.3 Japan

Analog systems

The analog systems in Japan work at 800 MHz and are outlined in Table 9.5.

Personal Digital Cellular

The Japanese standard for digital mobile radio is called *personal digital cellular* (PDC), although it has been referred to as JDC in the past. Like many of the digital systems today it has two variants differentiated by their operating frequency. They are PDC800 and PDC1500 operating in the 800- and 1,500-MHz bands. Unlike the analog systems in use in Japan, there will be only one digital standard (albeit with two operating bands). The aim is that

Table 9.5
Analog Mobile Phone Systems in Japan

Technology	MCS-L1	MCS-L2	JTACS	NTACS
Launched	1979	1988	1989	1991
Similar to	–	–	TACS	NAMPS

Source: [13].

this standard will offer all the ISDN-like features that one would expect from a digital mobile network. The voice coder uses a similar technology to that used in the American digital cellular systems, that is, *vector sum excited linear prediction* (VSELP). The fax transmission rate is limited to 4.8 Kbps.

An enhancement to the PDC standard will allow the use of a smart card to personalize the phone. Unlike the card used in GSM, this card can be removed from the phone and the phone will still operate. This opens up the possibility of manufacturing a multipurpose card (e.g., a card that can also be used in public phones, in public transportation, or in banking machines [15]).

9.5.4 Cordless Telephony

Cordless telephony standards are not only used in cordless phones, they are used in wireless PBXs and public telepoint services and wireless local loop. Cordless telephony standards have such a wide variety of applications because they only specify what is known as the *air interface* (i.e., the transmission between the phone and the base station). Details about network operation are up to the network designer or the PBX manufacturer.

Telepoint

Telepoint refers to a public network based on a cordless standard. These networks differ from cellular networks in the following ways:

- They only support pedestrian speeds, as opposed to vehicle speeds.
- They use a substantially higher bit rate (32 Kbps) for voice, allowing the production of cheaper and lighter handsets with longer battery life (talk time) and good speech quality.
- Cells are much smaller, allowing greater subscriber density but making nationwide coverage very costly. Service is usually only available in densely populated areas such as city centers, railway stations, and shopping complexes.
- The base stations can be mounted on telephone poles, on the sides of buildings, or in underground parking lots, avoiding the need for planning permission in most cases.
- Some telepoint services do not allow incoming calls. In the French Bi-Bop system, you must register with a base station if you wish to receive calls in a particular place.

For the above reasons, telepoint services cannot compete head to head with cellular services. Telepoint failed in the United Kingdom (after an investment of £120 million by four operators). The U.K. system only allowed out-going calls. The more successful telepoint services compete on the basis of lower tariffs and by making the reception of incoming calls as simple as possible [5].

For those who need the advantages of cordless and the coverage of cellular technology, the solution is a dual-mode cordless/cellular phone. One envisaged scenario is a phone that works on the office PBX using a cordless standard and will register itself with a cellular network, such as a GSM network, once you leave the building.

Cordless Standards

CT1. CT1 is an analog cordless phone standard and is used primarily in the consumer market. Many digital standards have been developed recently and can be used in wireless PBXs and public telepoint services.

CT2. The CT2 standard was developed in the United Kingdom for use in both public telepoint networks and domestic cordless phones. Initial proprietary CT2 specifications were replaced by *common air interface* (CAI) in 1989. The European Standards body, ETSI, adopted the CAI as an interim specification in 1992. The largest CT2 based services include: France (BiBop, 60,000 users), Hong Kong (170,000 users), and China (85,000 users) [5].

With the correct type of CT2 phone, data can be transmitted via an ordinary PSTN modem at speeds of 2,400 bps and sometimes better. In addition, 32 Kbps is possible using a data adapter provided both the phone and the network support this feature.

Digital Enhanced Cordless Telecommunications (DECT). The DECT specification was completed in 1992 and the first products appeared in 1994. Like CT2, DECT uses 32-Kbps channels for speech. There are two main distinctions between DECT and CT2:

- DECT channels can be easily aggregated to deliver ISDN speeds or greater.
- DECT has 120 voice channels, while CT2 has only 40. (In addition, there is a proposal to increase the number of channels in DECT.)

These advantages, coupled with the *exclusive* reservation of the 1,880- to 1,900-MHz band for DECT in all European Union countries, make DECT the most interesting cordless standard in Europe at present. Its primary applications so far are cordless PBXs and fixed radio access or wireless local loop (see Section 5.4.2).

Both DECT and CT2 use a technique called *dynamic channel assignment* (DCA), whereby the most suitable channel for each call is chosen by the equipment at call setup. This means that no frequency planning is required before installing base stations. This is a big advantage when deploying a cordless PBX in a large building.

Personal Handyphone System (PHS) [16].   An advanced cordless/telepoint system was launched in Japan in July 1995 and is expected to become available in most of the larger Japanese cities within a few years [15]. PHS phones can operate in three modes: as a cordless handset at home, as a cordless extension to a PBX in the office, and as a telepoint phone on the public network. Incoming calls and handoff between base stations are fully supported.

Personal Access Communications Services (PACS) [17].   PACS is a North American industry standard for cordless/telepoint applications. Adopted in 1995 and based on PHS and an earlier American standard called WACS, PACS is designed to work in the 1,900-Mhz PCS band (Section 9.5.2).

9.5.5 Mobile Data Networks [18]

While cellular networks are quite capable of handling data transmission, they currently have two drawbacks that can be overcome through the use of *mobile data* networks. These drawbacks are:

- Calls are charged on the basis of call duration rather than the volume of data transmitted.
- Call setup takes about 30 seconds, 20 seconds of which may be charged for.

Mobile data networks use packet-switching techniques to achieve better efficiencies when the data transmissions are bursty in nature (as described in Sections 7.3.2 and 7.3.3). This makes mobile data networks a good choice for applications such as database access, short email messages, and short file transfers. A packet service (GPRS) has been recently specified for GSM and promises most of the advantages of the mobile data networks discussed here. In addition it will have the advantage of integration with GSM voice services.

Typical Applications

Mobile data networks are typically used for dispatching jobs to service engineers and delivery staff. A service engineer, for example, might start work in the morning by logging onto the dispatch system and downloading a list of jobs. Without mobile data, he would have to travel to the office to pick up this list. As each job is completed, he can log a completion report directly onto the system, avoiding any time consuming paperwork. Alternatively, jobs can be dispatched one by one as the completion report for the previous job is filed.

The mobile data network can also be used to allow an engineer to consult technical documentation or client contracts stored on the office computer system. Some mobile sales staff use mobile data terminals to check if items are in stock and to place orders online thus offering faster delivery times to their customers.

Network Options

In the United States, there are three main options to choose from:

* RAM Mobile Data;
* Ardis;
* CDPD.

Of these, the first two have been in the market the longest and claim to provide coverage to 90% of the business population. This is lower than cellular, which covers 96% of the U.S. population [19]. Ardis and RAM do not, however, depend on roaming agreements for coast-to-coast coverage. Special roaming agreements now exist between some of these terrestrial-based systems and satellite-based systems, which effectively gives 100% coverage.

All three options use *store-and-forward* transmission to ensure that data is not lost if coverage is lost as you drive under a bridge or through a tunnel. Ardis, for example, will store messages for up to 72 hours.

Mobitex and RAM Data Network

The RAM Data Network in the United States is based on a proprietary system called Mobitex, which was developed by Ericsson. The radio link operates at 8 Kbps, but the actual throughput is about half this. Mobitex is also used in parts of Europe and Australia. There are a few roaming agreements in place within Europe. The RAM network in the United States is under the one network operator (carrier) and thus gives coast-to-coast service.

Mobitex-based networks allow communication between two mobile users or between mobile users and a fixed computer. Fixed computers are connected to the network via a wire-line link. This link would typically use the X.25, TCP/IP, or SNA data protocol (see Chapter 7). This fixed link can be direct or via another data network as shown in Figure 9.9. It allows businesses to attach a

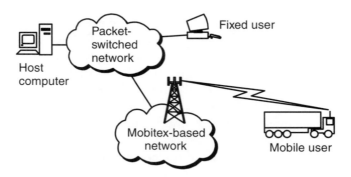

Figure 9.9 Mobitex allows communication between mobile users and fixed hosts.

host computer via a more reliable and, if need be, a faster connection than the radio connections used by the mobile users.

Ardis

The Ardis network is based on a proprietary protocol developed by Motorola and IBM. It operates at a speed of 4.8 Kbps (19.2 Kbps in some cities) but, as with all packet radio networks, the actual throughput is about half this.

Ardis is accessed via a portable computer or a PDA with an Ardis-specific PCMCIA modem. Services are as follows:

- Ardis-to-Ardis messaging—sender pays;
- PSTN-to-Ardis paging via touch-tone phone or operator—receiver pays;
- Ardis to fax;
- Internet to/from Ardis.

CDPD

This is a packet-switched network that uses idle capacity in AMPS-based cellular networks to transmit packets of data. CDPD offers the following features:

- It supports a bit rate of 19.2 Kbps (about 10 Kbps after overhead).
- There is very low call setup delay; packets are transmitted into the network with an IP address in the header and this address routes the packet to its destination.
- CDPD modems cost around $400 in 1996 [19].

Rollout of CDPD in the United States began in 1994. It is not very widely deployed (1996), and roaming agreements are required between the operators if true coast-to-coast coverage is to be achieved.

9.5.6 Satellite-Based Mobile Communications

One of the problems with cellular networks today is the prohibitive cost of providing coverage in sparsely populated regions. Globally, less than 15% of the world's land mass is covered by cellular services [20]. The most economic solution to this problem at present is the provision of satellite-based mobile services.

A second major problem with today's cellular phone systems is the impossibility of true global roaming. A satellite-based system can solve this problem because one system can cover the globe. To keep call costs down, dual-mode phones can be used, which will make a call over a cellular network rather than the satellite if within coverage of the cellular network.

Mobile Satellite Services

Existing (1997) mobile satellite services for voice communication such as the international Inmarsat-M service and the U.S.-based SKYCELL service are based on geostationary satellites. The phones are about the size of a notebook computer or larger and have a directional antenna that must be pointed at the satellite.

Recently there have been many initiatives aimed at providing mobile communication for *handheld* phones using satellites in lower orbits—so called *low Earth orbit* (LEO), *medium Earth orbit* (MEO), and *intermediate circular orbit* (ICO). These orbits require more satellites to give global coverage. However, they do allow for handheld phones, and the propagation delay over the communication path is reduced.

Some of these proposals are listed in Table 9.6. All of the systems operate at frequencies between 1.6 and 2.5 GHz, even though there is very little bandwidth available at these frequencies. The problem with higher frequencies is that they require directional antennas, thus making the handset more difficult to use.

All systems support dual-mode phones (i.e., the phone will be capable of connecting to a land-based service such as GSM or AMPS whenever it is available). This should reduce the cost of ownership and will allow the phone to be used indoors in many instances.

These services will not necessarily be available in every country because the operators must get regulatory approval locally, and there is no guarantee that this will actually happen. Handsets are expected to be more expensive than their cellular counterparts and are not expected to work indoors in satellite mode.

Table 9.6
Satellite-Based Mobile Phone Systems

System	Iridium (Motorola)	Globalstar (Loral Qualcom)	Odyssey (TRW)	Inmarsat P (ICO)
Expected in service	1998	1998	1998	2000
Coverage	Global	Limited Global	Global	Global
Number of satellites	66	48	12	10
Altitude (km)	780	1,387	10,355	10,355

Source: [21] with permission; see also [22].

Another limitation on these systems is that they will support a relatively low bit rate in satellite mode (e.g., Iridium will only support a user bit rate of 2.4 Kbps for voice, fax, and data [23]). This will limit the voice quality and will make the sending of fax messages both expensive and time consuming.

9.6 Paging

Paging systems began as national, one-way, tone-only systems. Voice and text pagers followed, and the latest paging systems to emerge are international and/or two-way. Pagers are cheaper to own and operate than mobile phones and thus have quite an appeal in the consumer market. They are also a very useful business tool. Pagers are much smaller than mobile phones and the batteries last much longer—wristwatch pagers have been available since 1991.

Pagers are generally used as a means of asking somebody to call back or to call his or her voice mail to listen to a message. A telephone number can be sent to a pager by dialing the pager number and then keying in the number on the telephone keypad.

Many pagers can receive short text messages. One way of sending such a message is to call a human operator, speak the message to them, and they will then send the message. The message can also be originated from a PC via an email system, provided the paging network supports this feature.

The biggest drawback with one-way paging systems is the lack of confirmation of receipt of the message. Messages can be missed if the pager is switched off or is outside the coverage area, and they will not be retransmitted when the pager is switched back on. Two-way paging systems can overcome this problem. Two-way paging was introduced the United States in 1995.

9.6.1 European Radio Messaging Service

European radio messaging service (ERMES) is a European paging standard adopted in 1992 and later adopted by the ITU as an international standard. Most paging networks however still conform to the older *post office code standardization advisory group* (POCSAG) standard. ERMES operates at 6 Kbps and allows for international roaming [24].

9.6.2 Satellite-Based Paging Systems

Most of today's paging networks are based on a terrestrial distribution system involving base stations and antennas. Satellite-based systems offer better coverage, particularly in remote areas and in mountainous regions.

References

[1] Minges, Michael, and Tim Kelly, *World Telecommunication Development Report 1996/97: Trade in Telecommunications Executive Summary,* International Telecommunications Union, http://www.itu.int/ti/publications/world/summary/wtdr96.htm, February 1997.

[2] Calhoun, George, *Digital Cellular Radio,* Norwood, MA: Artech House, 1988.

[3] Office of Telecommunications, International Trade Administration, U.S. Department of Commerce. "World Cellular Market Table," www.ita.doc.gov/industry/tai/telecom/telecom.html, 1997.

[4] Ericsson web pages, http://www.ericsson.com/Reports/9month96.html.

[5] Hadden, Alan David, *Personal Communications Networks: Practical Implementation,* Norwood, MA: Artech House, 1995.

[6] GSM MOU Association website, http://www.gsmworld.com.

[7] Mouly, Michael, and Marie-Bernadette Pautet, *The GSM System for Mobile Communications,* Palaiseau, France, published by authors, 1992.

[8] Lattimore, P., "Small steps to a big cell," *What Cellphone,* Feb/March 1995.

[9] Guest, Tim, "A bird in the hand," *Mobile Europe,* March 1996.

[10] Gareiss, Ross, "Wireless Data—More Than Wishful Thinking," *Data Communications International,* March 21, 1995.

[11] Rysavy, Peter, and Craig J. Mathias, "Digital Cellular Networks—On the Road to PCS," *Network Computing,* Feb. 1996.

[12] Smith, A., "There's life in the old battery yet," *Mobile Europe,* July/August 1995.

[13] Isaksson, M., A. Bacon, and E. Grönstad, "D-AMPS 1900—The Dual-Band Personal Communications System," *Ericsson Review,* No. 2, 1995.

[14] Balston, D. M., and R. C. V. Macario (eds.), *Cellular Radio Systems,* Norwood, MA: Artech House, 1993.

[15] Lehmann, Yves, "Japan on the Move," *Telecommunications™,* International edition, March 1995.

[16] Mansfield, Simon, "PHS: A Pocket Full of Talk," *Internet Access Center K.K.,* http://www.jpn.co.jp/dec95/jp3.html, Nov. 20, 1995.

[17] PACS Providers Forum website, http://www.pacs.org.

[18] Gareiss, Robin, "The Business Case: Don't Have a Plan? Don't Cut That Cable," *Data Communications International,* May 21, 1996.

[19] Ross, T., Private Communication, MTA-EMCI, Inc., Washington, DC, June 1995, quoted in Ronald H. Brown, and Larry Irving, *NTIA Special Publication 95-33: Survey of Rural Information Infrastructure Technologies,* Communications and Information, and Administrator, National Telecommunications and Information Administration U.S. Department of Commerce, http://ntia.its.bldrdoc.gov/its/spectrum/rural/, September 1995.

[20] ICO Global Communications, "Background to ICO," http://www.i-co.co.uk, 1995.

[21] Johannsen, Klaus G., "Mobile P-Service Satellite System Comparison," *International Journal of Satellite Communications,* Vol. 13, 1995.

[22] Tor, E. Wisløff, "Big LEO Overview," http://www.idt.unit.no/~torwi/synopsis.html.

[23] "Global Wireless in the Palm of Your Hand," Iridium web pages, http://www.iridium.com, 1997.

[24] Claydon, Lieselotte, and Lars Gandils, "ERMES Set for Take-Off," *Telecommunications™,* International edition, March 1995.

10

Value-Added Services

10.1 Value-Added Services

The term *value-added services* is almost impossible to define for a book such as this. The term means many different things in different markets. In many cases, the use of the term is synonymous with *unreserved* services. This is because until recently, the network operators in most European countries and the local operators in the United States had a legal monopoly on basic services, whereas value-added services were open to competition. As time progresses, however, this usage becomes invalid as services once considered basic in the past become open to competition [1].

Therefore, we will not attempt to define *value-added services*. We have merely chosen the term as the title of this chapter, in which we discuss a range of telecommunications services commonly referred to as value-added services.

10.2 Information-Access Services

This is a group of services that provides users with a wide variety of information for various purposes. Some services are primarily used by residential users; others are targeted at business users. In most cases, the service is accessed via the telephone network, although for a number of cases, the service is provided over a private network with access via leased lines.

10.2.1 Freephone

Freephone describes the service in which the cost of calls is paid by the party receiving the call. The person making the call is not charged, hence the general term *freephone*. These services have been around for some time now, and are commonly referred to by the prefix used to access the service (e.g., "800" or "888" in the U.S.; "0800" in the U.K.). Most operators also provide *enhanced freephone* services, which include additional features, such as routing the call to different locations based on the origin of the call, the time of day, and other factors.

The benefit of freephone service is that it encourages incoming calls; therefore, it is most often used by companies for sales and customer-support lines. In this way, the company pays for the customer support, but may benefit in the long run by increased customer loyalty.

A variation of the freephone concept is *low-cost calls,* where the calling party pays only a small proportion of the price of the call, while the called party pays for the remainder. Typically, the calling party pays the same charge as for a local call. The benefits of this service are similar to freephone, in that it encourages incoming calls. However, as the caller has to pay a small fee, it can reduce the number of unwanted calls.

Until recently, one of the major disadvantages of value-added services such as freephone was that they were generally restricted to calls within the same network. In most cases, this meant that they were limited to national usage.

There are three approaches to providing international freephone services.

1. The company subscribes to a different national freephone number in each country. Calls to these numbers are then redirected to the call handling center or centers. This has the disadvantage that any promotional material must contain the different freephone numbers for each country.

2. There are agreements between a number of network operators to allow incoming international access to freephone numbers. For example, to dial the U.S. freephone number 800 xxx–xxxx, a caller can dial +1 800 xxx–xxxx. Before being connected, however, a message is normally played warning the caller that the call will be charged at international direct-dialing rates. In short, foreign callers can dial the freephone number but do not get the free service.

3. *Universal international freephone number* has been recently introduced, following the publication of ITU-T Recommendation E.169. This allows one number to be used anywhere in the world, the format of the number being +800-GSN, where GSN is the company's particular eight-digit *global subscriber number*. The company, however, still has to subscribe to the services of the network operators

in each of the individual countries. The big advantage is that a genuine freephone service is provided in all the relevant countries, and that only the one freephone number is required for promotional material.

10.2.2 Videotext

Videotext is a particular standard for access to online databases, originally adopted by the PTTs. Its unique selling point is that a standard TV can be used to view the data, with access via an ordinary telephone line. In addition to straightforward database access, the standard also allows for a degree of *interactivity*, for example a user having viewed an airline's flight times can then book a seat. The information provided on a videotext network is provided by a large number of sources. Some provide information free of charge, while others charge for the service. The fees are generally collected by the network operator, which then passes on the revenues to the information providers.

However, videotext has not been successful in most countries—in fact, in the United Kingdom, the Netherlands, and Sweden, the PTTs have sold or closed down their videotext operations. The big exception to this is the French market, where *Minitel* has been very successful, having benefited from substantial government backing. The electronic telephone directory (*l'Annuaire Electronique*) is by far the most popular service. There are over 20,000 services on offer and 6.5 million Minitel terminals in France [2].

10.2.3 Premium-Rate Services

Premium-rate services or audiotext or "900" services (U.S.) refer to the provision of "information" over a telephone call. Unlike freephone, the cost of the call is paid by the calling party, usually at a significantly higher rate than normal (hence the commonly used term *premium rate*). The revenue from the calls is collected by the network operator and a proportion is paid over to the service provider.

The benefit of premium-rate services is that they generate revenue from incoming calls. Thus, it is commonly used for information services such as sports results and weather forecasts. It is also used by some companies to provide customer-support lines—in this instance the customer pays for the support through the call charges.

Premium-rate services are probably the most controversial value-added service, due to a large number of "adult oriented" services in many countries.

10.2.4 Online Databases

Online databases provide access to a wide variety of historical and real-time information.

The largest share of the online databases market is for real-time financial and business information, such as that provided by companies such as Reuters and Telerate. These cover such *real-time* information as foreign exchange, interest rates, equities, commodities, energy market spot prices, shipping, municipal bonds, mortgage markets, pending legislation, legal convictions, historical data, and analyses of company and business information. Databases containing more general information are also available, such as those provided via CompuServe and the Internet.

The equipment required to access online databases varies depending on the particular service. Most databases, however, are accessible using a PC and modem over a telephone line or ISDN access.

10.3 Messaging and Transactional Services

These services involve the transfer of information according to standardized protocols.

10.3.1 Electronic Mail

Electronic mail refers to the delivery of information by electronic means. It is often described as the telecommunications equivalent of the postal service. The term *email* generally refers to the *system* used for the delivery of the information; the actual content and structure of the information is not specified. In fact, *electronic data interchange* (EDI) and *electronic funds transfer* (EFT) (covered later in this chapter) are examples of particular types of information that can be transferred over an email system.

Advantages of Email

Email is faster than the normal postal system, with delivery times of the order of several minutes. However, where delivery time is critical, sending a fax message may be a better option. This is mainly because the recipient of email must, in many cases, first check their mail box before they are aware of any incoming messages—some users may only check their mail on a daily or even weekly basis! As well as this, the delivery time for email is variable; delivery times of over an hour can be expected during peak times on some networks.

Another advantage is that most business documents are prepared on computer systems by the sender and stored on computer systems by the receiver—the use of email for such documents has obvious benefits.

Email also has advantages for sending messages that would otherwise have been communicated over a telephone call, as the recipient does not need to be available at the time of sending the message. Furthermore, when the recipient retrieves the message, they have a written record of the message.

It is also possible to configure an email system so that the users can access their mail boxes remotely. Thus, messages can be received when away from the office—a big advantage over the normal postal system. Many email services on the Internet can be accessed from anywhere in the world.

Additional features which may be provided by an email service include:

- Multiple addressing: the same message may be sent to more than one recipient at the same time.

- Delivery and/or receipt notification: The sender can be notified of successful/failed delivery to the recipient's mailbox (delivery notification) and/or whether the recipient has retrieved the message from the mailbox (receipt notification).

- Delivery options: The time at which messages will be delivered can be specified. Recipients can also redirect messages to an alternative recipient, while on vacation for example.

- Priority: Messages can be marked as *urgent, normal,* or *nonurgent.* This is used to signal to the message-handling system the required speed of delivery. If the network is congested, the *urgent* messages will be handled first.

- Attachments: In addition to the main text, other files can be attached to the message (e.g., images, databases, graphs) (see *multipurpose Internet mail extensions* [MIME] later in this chapter).

- Message Storage: Messages can be stored for retrieval at a later time.

- Conversion: Email text messages can be converted to fax, teletext, telex, and vice versa.

As email developed, a lot of different protocols emerged, most of which were incompatible with each other. This led in the 1980s to the development of the X.400 recommendations by the CCITT, intended as a worldwide standard [3]. However, with the meteoric rise in recent years of the Internet, its email standard called *simple mail transfer protocol* (SMTP) is now more widely used.

Different email systems are interconnected through *gateways,* thus it is possible for a user of an SMTP system to send messages to an X.400 user, and vice versa.

X.400 Versus SMTP [4]

Both X.400 and SMTP handle the basic functions of email. If additional functions are required, careful consideration should be given when deciding which of the two systems you should use.

- Cost: X.400 mail systems are considerably more expensive to subscribe to and use than Internet mail systems. In fact, some Internet mail systems are totally free for the user!

- Ease of use: Internet mail systems are usually much simpler to use than X.400 systems. The most obvious example is in the addressing of messages. My own Internet address is: *pwalsh@telecom.ie,* whereas my X.400 address would be: */c=ie/a=eirmail400/p=telecom/o=tex400/s= Walsh/g=Philip.*

- Features: X.400 was designed to include a lot of features from the outset, whereas SMTP was designed simply for the transfer of text messages. However, extensions to the SMTP protocol such as *multipurpose Internet mail extensions* (MIME) now also provide the more important features. One of the main features that MIME supports is the attachment of various types of file to an email message in a standardized way. Both systems now allow attachments, delivery notification, and message-storage functions. X.400 also provides priority, delivery options, reliable transfer service, conversion of messages, and receipt notifications.

- Security: X.400 is apparently more secure than Internet mail. However, although a lot of security features have been *defined* for X.400, few of them are in widespread use. These features include: message integrity, message confidentiality, nonrepudiation of origin, and nonrepudiation of delivery. With regard to security for email on the Internet, the main concerns relate to encryption of the message and authentication of the sender (I have even heard of messages being received from "God@Heaven.org"). A number of different security methods have been developed including *Pretty Good Privacy* (PGP) and secure MIME (S/MIME). These systems do not offer as many security services as a full X.400 implementation: they include message encryption and authentication but do not include proof-of-delivery or nonrepudiation services. A protocol known as *message security protocol* (MSP) used in

X.400 provides nonrepudiation of receipt and has recently been adapted for use with Internet email [5].

- Respectability: X.400 was created by reputable standards bodies—the CCITT (since taken over by ITU-T) and the ISO. Internet mail was defined by the *Internet Engineering Task Force* (IETF), which is a volunteer organization. However, nowadays the IETF is recognized as a valid standardization body by both the ISO and ITU. Because of problems with the authentication of Internet mail (i.e., it is relatively easy to send messages posing as somebody else), only X.400 messages are currently recognized in legal cases. This may change in the future as the security features of Internet mail improve.

- Directory services: The X.400 standard has a related standard called X.500 for an integrated directory service. However, although the standard exists, as yet there is no large-scale X.500 directories in existence. The Internet mail system does not define a standard for directory services. A number of directory services do exist, but as yet there is no large-scale directory.

- Number of users: Internet mail has considerably more users than X.400. X.400 users tend to be in large companies whereas Internet users tend to be in large and small businesses or home users. It is possible to send messages between X.400 and Internet mail users through gateways.

10.3.2 Electronic Data Interchange

EDI is the exchange of formatted business data between computer systems. It is similar to email but has a number of important differences. Email refers to the exchange of messages between people, whereas EDI refers to the transfer of data between computer systems. Because EDI messages are received by computer systems, the format of the data messages must *exactly* conform to agreed standards. In addition, there is increased importance on the *integrity* of the data. For example, if an email message is slightly corrupted it may be still intelligible to the receiver, since the human brain is able to fill in the gaps. EDI information must be received exactly as transmitted, since computer systems do not have the same degree of intelligence (yet!).

At present, most EDI transactions are made over *value-added networks* (VANs). Users usually dial into the VAN to send and/or receive EDI messages. The VAN acts as a trusted third party and guarantees delivery and integrity of the transactions. The VAN also authenticates the users though the use of logon-IDs and passwords so that recipients of messages can be assured that transactions are coming from the person or entity that they purport to have come from. Finally, the VAN can provide various notifications to the sender as to the progress of the transaction, such as *mail box pickup notification,* which informs the sender that the recipient has downloaded the transaction from the network [6].

Proprietary mechanisms are now available for transmitting EDI messages securely over the Internet (or any IP network). A standards-based solution is being drafted (1997) [6].

There are two sets of standards for EDI:

1. *Formatting standards* [7]: These define the way in which the information contained within an EDI message are formatted. EDIFACT is the most important standard, being a United Nations–sponsored standard for EDI introduced in 1987. Older standards still in use include X.12, which is an ANSI standard used in the United States, and UNTDI, which is a United Nations standard used in Europe. The TRADACOMS standard used in the United Kingdom is one particular national version of UNTDI.

2. *Transfer standards:* These define how the information is transported from end to end. The main systems in use for wide area EDI are X.400 systems. The X.435 standard is particularly defined for the transfer of EDI messages and includes mandatory notifications and specific headers containing EDI information. However, because X.435 is relatively recent, many users use the X.420 (normal email) standard for transferring EDI messages.

In 1995 a definition of a MIME EDI standard was published as RFC1767 to enable the transfer of EDI messages over the Internet mail system. RFC1767 does not include important functions such as integrity, confidentiality, and nonrepudiation of origin and receipt. An additional standard is being drafted [7] to rectify this.

The main benefits of EDI are increased productivity and reduction in the amount of errors, a result of the data not having to be keyed in manually by a human operator. This results in savings, both through the reduction in the cost of handling the information (most estimates reckon about a 25% saving [7]), and also because company's processes can work faster—therefore there is less capital tied up in stock (inventory) and cash flow is improved.

There are also intangible benefits for the company, the improved processes leading to increased customer satisfaction, "suppliers" becoming "business partners" and an overall better competitive edge.

10.3.3 Electronic Funds Transfer

EFT is a particular form of EDI, where the documents being transferred contain money. However, some in the banking/financial community claim that EFT is not EDI, as it came from a different group and has different formats and rules. EFT refers to the transfer of funds between banks, financial institutions, stock exchange trading systems, and similar organizations. The main network used for EFT in the world is an X.400-based network run by SWIFT. SWIFT is a closed user group owned by a large number of banks throughout the world. These banks are therefore the main customers (though not the only ones) of the network [1].

10.3.4 Electronic Funds Transfer at Point of Sale

Electronic funds transfer at point of sale (EFTPOS) is another particular form of EDI. Like traditional EFT, the documents contain money but in this case the money is transferred from a customer's account to a retailer's account and, by definition, the transfer takes place at the point of sale.

Payment is by means of a credit card, debit card, or prepaid debit card. The best systems are *online,* in which case after the card is "swiped" through the card reader and, if necessary, the customer enters a PIN code, the customer's account is interrogated immediately for authorization. If the customer has sufficient funds or credit, then the funds are transferred to the retailer's account automatically. There are a number of *offline* systems, in which case the transfer of funds is done on a batch basis, usually at off-peak times.

The advantages of EFTPOS for a retailing company are:

- Reduction in money lost due to bounced checks or credit card fraud;
- Reduction in amount of money on the premises; therefore, less of a security risk;
- Reduction in the amount of paperwork compared with traditional credit card payment, checks, and cash;
- More convenience for customers than other payment methods.

10.3.5 Voice Mail

Voice mail essentially provides the same features as a telephone answering machine except that the equipment for storing the messages is located on the network. This has a cost advantage because the equipment is shared by many users.

It also has the advantage that the user does not need to be concerned about maintaining and upgrading the equipment.

Because voice mail can be accessed from *any* telephone, the voice-mail user does not actually require a phone line! This particular use of voice mail is often called *virtual telephony*. This is useful for families, where each member of the family can have their own mail box accessible via the family phone line, or where one member of the family uses the mail box for conducting business related to community or sports clubs.

Public voice mail is generally aimed at domestic users, but has some advantages for small businesses or people who work from home. It is also an important feature of many mobile networks, where messages can be received for users who are out of coverage or have their units powered off. For large companies, however, the voice-mail service is generally provided via the PBX.

10.3.6 Fax Store and Forward

Fax store and forward is essentially a voice-mail service for fax machines. In fact, in many cases the fax store-and-forward service is provided by the same equipment as the voice-mail service, the equipment simply identifying a fax call by the characteristic tones sent by a fax when setting up a call.

The advantage of this service is the same as for voice mail, in that incoming messages are not lost due to the fax machine being busy or temporarily out of service.

One disadvantage, however, is that some systems can introduce a considerable delay between the time when the message is sent by the sender to the system and the time when it is eventually forwarded to the recipient.

10.4 Virtual Private Networks

VPNs use a public switched network to imitate the features and facilities of a private network. VPNs are available on data networks, but the major area of interest is in VPNs implemented on the PSTN/ISDN network. In this case, the network equipment is the exchanges and transmission network of the PSTN/ISDN, simply with additional software added. VPNs are often implemented using the *intelligent network* architecture discussed in Section 10.5.

The main operational features of a VPN service are:

- Private numbering plan: *On-net* calls, calls between two users on the same VPN, are made by dialing short *private numbers,* typically four to six digits long.

- Charges: Charging is somewhere between leased lines and standard PSTN lines, with relatively high annual rental charges and relatively low call charges.

VPNs originated in the United States, being operated by the long-distance carriers such as AT&T, Sprint, and MCI. In part, the success of VPNs in the United States may be because it offered a form of competition between the long-distance carriers and the regional telephone company at a time when the carriers were prohibited from offering a local service.

Development of VPNs elsewhere has been slower. Until recently most European countries merely offered international VPN services in collaboration with the American carriers. In many cases, incompatibilities between the exchanges mean that a very limited feature set is possible, usually only abbreviated dialing and volume discounts.

This is changing, with many European operators now offering (or planning to offer) sophisticated national VPNs.

10.4.1 Benefits of VPNs

Costs

There are two ways in which VPNs offer cost advantages to the customer.

1. Call charges: In most cases, VPN calls are charged at a lower rate than standard PSTN/ISDN rates. In addition, volume discounts normally apply to all traffic on the VPN, including traffic from low-volume sites that may not normally qualify for volume discounts.
2. Overheads: As the VPN is managed and maintained by the network operator, the user does not have to provide for these overhead costs. There can be a considerable saving over the costs of managing and maintaining a private network.

Flexibility

As the access to the VPN is via standard PSTN/ISDN lines, new sites can be brought *on-net* much more rapidly than connecting new sites using leased lines.

As the VPN has, theoretically, the entire capacity of the PSTN/ISDN available, the VPN can be rapidly reconfigured to suit changing demands.

Another advantage of using the PSTN/ISDN as the platform for a VPN is that in most cases *remote access* is available. This means that staff who work outside of the normal office environment (e.g., sales staff or field engineers) can dial into the VPN using an access code. This means that they can make calls from any phone, which will then be charged to their VPN account.

Reliability and Availability

The VPN has the same (relatively high) reliability and availability as the PSTN/ISDN. It can be expensive to provide the same reliability in a private network by providing redundant nodes and circuit rerouting.

Features

The VPN service can also offer a range of features in addition to the basic VPN service. For example, the following features are commonly offered:

- Call diversion: Calls can be diverted on "busy" or "no-reply." In some cases, call diversion can be user controlled.
- Time-dependent routing: Calls can be routed to various destinations based on the time of the day or the day of the week.
- Origin-dependent routing: Calls can be routed depending on the origin of the call, usually to route calls to a local office.
- Call distribution: Calls can be distributed over a number of lines, usually for calls incoming to an operator.
- Customized announcements in different languages, depending on the origin of the call.
- Call screening: Users can be barred from making calls, such as international calls or premium rate calls.
- Speed dialing: Short dialing codes (e.g., one to four digits) for non-VPN or off-net numbers.

Management and Maintenance

The management and maintenance of the VPN remains the responsibility of the network provider. Usually sophisticated management and statistical reports are provided to the VPN customer company to ensure that the optimum service is provided. Billing and charging is usually more sophisticated and flexible than standard PSTN billing, and usually a number of accounts can be provided for each VPN to accurately apportion costs to business units.

10.4.2 Centrex

Centrex is a service provided by a public network operator where a segment of a public exchange (central office) is configured to behave like a PBX (see Chapter 3). The client rents a number of centrex lines instead of purchasing a PBX. All the typical PBX features, such as short-code dialing between lines, call transfer, and call waiting, are available on these lines. DID will also be available as standard. The centrex and VPN concepts can be combined for multisite solutions. Thus, a VPN links a number of sites served by centrex lines.

There is very little capital outlay with centrex, compared to the option of purchasing and installing a PBX on your premises. Thus, centrex might be a very suitable solution for a temporary site or as a means of buying time as you go to tender for a PBX system. One point to remember here is that price comparisons between centrex and a PBX are complex because there are very different cash flow profiles in each case.

10.5 Intelligent Network

The term *intelligent network* (IN) is applied to new developments of the PSTN/ISDN network. There are many different interpretations of IN; in fact one of the major obstacles to studying the subject of INs is the vast number of definitions of the term. Essentially intelligent networking implies the following:

- The use of increased computer power to control the switches in the network;
- The provision of advanced value-added services rapidly and efficiently.

The original dream of IN was that new services could be developed, tested, and made commercially available within a few days. However, the experience to date is that new services require about six weeks of further development before being commercially available. This is still considerably faster than the time scales involved when implementing new services using traditional networks.

In many ways, the old manually operated telephone networks were INs, where the intelligence resided in the brains of the telephone operators. For example, I'm reminded of the story of a friend who called her mother in a small rural town. On asking the operator to be connected to the particular telephone number, the operator replied, "Your mother has gone out to the hairdresser's. She passed my window just five minutes ago. Would you like me to put you through to her at the hairdresser's?" Modern INs can be seen as a way of reintroducing the operator's intelligence back into the network; however, in the modern case the intelligence resides on a computer.

Typical services provided by intelligent networks are:

- Freephone services;
- Premium-rate services;
- Virtual private networks;
- Charge card and credit card calling: The user can charge calls to a network provider's charge card or to a standard credit card. The service can be provided automatically or using the assistance of an operator;
- Televoting: By televoting, users can choose between a number of different options, each option having its own particular number. The user votes by dialing the number of his or her particular choice.

It must be explained that IN is not the only way in which these services can be implemented—many of the above services can be (and have been) implemented using traditional network architectures. Neither are they the only services that can be provided using IN—IN has been specifically proposed as a way in which new, previously undreamed of, services can be implemented easily and rapidly.

10.5.1 Operation of an IN

We will use the example of a freephone call to illustrate the operation of an IN. Other services obviously differ in their implementation but follow the same basic pattern.

Definition of the Service

The general freephone service logic is defined using standard *service-independent building blocks* (SIBs) and stored. When a new service provider subscribes to the service, the network operator loads the general freephone service logic and adds the particular details of the new freephone subscriber. The resultant *service script* is then downloaded to the *service control point* (SCP), and the service is now available throughout the network. The SCP can be a modified telephone exchange or a general-purpose computer connected to the PSTN/ISDN using common channel signaling.

Call Processing

When a service user dials the freephone number, the call will trigger a signaling message to the SCP. This message will typically include the number dialed and the calling party's number, category, and area code.

When the SCP receives the message, it analyzes the number dialed and as a result is directed to execute the service script for the particular freephone subscriber. In this case, the result of the service script will be the "normal" number of the freephone subscriber. This information is returned to the network in a signaling message.

On receipt of this message, the call will be connected through to the normal number of the freephone service provider as if the call was a normal call.

10.5.2 Advantages of IN

From the above description, a number of advantages of IN should become apparent:

- The *intelligence* is stored in one, or a small number of, SCPs. Thus, it is much easier to maintain the information, to introduce new subscribers to existing services, and to define entirely new services.

- The creation of service logic using SIBs means that new and powerful services can be defined, tested, and introduced much more rapidly than using traditional methods. New services can be defined and provided by equipment suppliers, network providers, or even third parties.

- By using sophisticated common channel signaling between the SCPs and the main exchanges in the network (now called *service switching points* [SSPs]), only the information needs to be transferred to and from the SCPs. Once all the information has been returned by the SCP to the network, the call can be routed using the most optimum routing methods. This can result in significant cost savings for the network operators.

In summary, INs do not provide anything really new for customers. But by enabling services to be developed far more rapidly and cheaply than traditional network architectures would allow, they allow services to be offered by network providers that might not have been economically viable in the past.

References

[1] Communications and Information Technology Research, *Value Added Services in Europe,* CIT Research, London, 1994.

[2] "Overview of Minitel in France," France Télécom Intelmatique, http://www.minitel. fr/English/Minitel/overview.html, 1995.

[3] Green, D. C., *Data Communication,* London: Longman Scientific & Technical, 1991.

[4]   Alvestrand, Harald T., "The Canonical Internet vs X.400 Debate," http://domen.uninett.
       no/~hta/x400/debate, 1996.

[5]   Levien, Raph, "Protecting Internet E-mail From Prying Eyes," *Data Communications International*, May 1996.

[6]   Shih, Chuck, Mats Jansson, Rik Drummond, and Lincoln Yarbrough, "EDIINT Functional Specification—Requirements for Inter-Operable Internet EDI," Internet Draft EDI-INT Working Group, http://www.imc.org/ietf-ediint, March 1996.

[7]   Reardon, Ray, *Future Networks: New Developments New Opportunities,* London: Blenheim
       Online, 1989.

11

Communication Equipment

11.1 Introduction

In this chapter we will look at modems, multiplexers, and video equipment. Many more pieces of equipment coming under the heading "communication equipment" have been dealt with in other chapters of the book.

11.2 Modems

Modems connect computers or other devices to a communication network. Their function is to convert the signals coming from the computer into signals suitable for the network. The most common type of modem is that used for connection to a PSTN line, used, for example, by those who have Internet access at home. Modems are also used by small businesses for dialup access to Internet, information, email, or EDI services. Most modems can be used for sending and receiving fax messages, while some modems can also be used for voice mail.

Modems used on the PSTN are called *voiceband* modems because they take the digital signal from the computer and convert it into audio tones suitable for transmission over the telephone network (i.e., within the voiceband).

Apart from their use on the PSTN, modems are also used on other analog communication links such as analog leased lines and analog access links to public data networks (Figure 11.1). In both cases the modems must be capable of *leased-line operation* because the circuit between the customer's premises and the public data network is technically equivalent to a leased line.

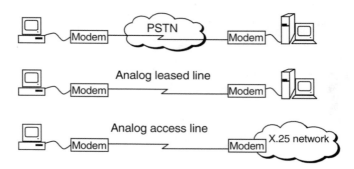

Figure 11.1 Analog modems are used on PSTN lines, analog leased lines, and analog access lines.

11.2.1 Fax-Modems

Most of today's dialup modems have the ability to send and receive faxes and are sometimes referred to as *fax-modems*. A fax-modem connected to a PC can be used to fax information stored on the PC. Anything you can print can be faxed. A very useful feature is the ability to send out *mailshots* by fax using a phone book stored on the computer (Figure 11.2). Software is also available that will allow you send a *mailmerge* document via the fax-modem, thus allowing each fax to be personalized with the recipient's name and other individual details.

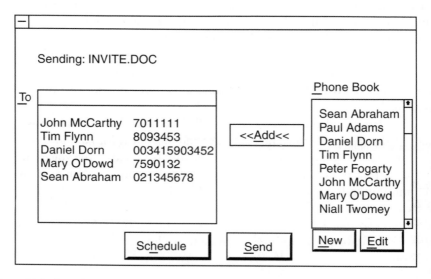

Figure 11.2 Preparing a mailshot for sending via a fax-modem using fax software on a PC.

Fax-modems are a convenient way of sending responses to customer queries, particularly when the answers (price lists, product descriptions) are already stored on the PC. It is often possible to have the response sent while still speaking on the phone to the customer.

It is also possible to use fax-modems in conjunction with automated response units or PC software that will allow you to create a fax-back service. Your clients can then call into this service from a fax machine and request particular documents to be faxed back to them by keying appropriate digits on the keypad.

11.2.2 Modem Standards

The majority of voiceband modems today conform to international standards published by the ITU-T. There are many such standards because modem technology has progressed over the years. Table 11.1 shows a small sample of these standards. V.32 *bis* and V.34 are by far the most popular standards today.

The advantage of these standards is that modems purchased from different manufacturers can interwork. In practice, unfortunately, interoperability problems between modems conforming to the same standards are quite common, particularly among modems that were manufactured within a couple of years of the introduction of the standard.

The standard numbers are arbitrary (i.e., there is no way of calculating the modem speed from the number). The *bis* after some numbers literally means "twice" (in Latin). In practice, *bis* means a new standard that is strongly related to the first. Similarly *ter* refers to a third new standard.

Table 11.1
Modem Standards From the ITU-T

Standard Number	Bits per Second	Suitable for
V.21	300	PSTN
V.22	1,200	PSTN or two-wire leased line
V.29	9,600	Four-wire leased line
V.32	9,600	PSTN or two-wire leased line
V.32 bis	14,400	PSTN or two-wire leased line
V.34	33,600	PSTN or two-wire leased line

11.2.3 Modem Speeds

The latest modem standard (V.34) specifies a modem that can operate at 33.6 Kbps over the telephone network. This speed is very close to the theoretical limit of what can be achieved over analog PSTN lines.

Modem manufacturers have recently developed modems that can work at a nominal 56 Kbps in one direction and at 33.6 Kbps in the other. The 56 Kbps will only work under the following conditions:

- One end of the connection is on a *digital* line (e.g., an ISDN line). The modem on the analog line receives data at up to 56 Kbps and transmits at up to 33.6 Kbps.

- The modems at both ends use the same chips (i.e., in the short term there is no internationally agreed standard for 56 Kbps).[1]

- The connection is perfect. In practice expect 40 to 50 Kbps.

If higher dialup speeds are required, then ISDN is required at both ends of the connection and *terminal adapters* are used instead of modems.

Line Impairments

The speed a modem will operate at depends on the capability of the modem and the quality of the connection. Noise on the line or some other impairment will tend to reduce the speed. Problems can be caused by poor wiring either on your premises or anywhere on the connection between the two modems.

It is quite common for modems to operate below their maximum speed, so that if a V.34 modem consistently connects at, say, 26.4 Kbps, there may be very little that can be done about it. The degradation may be caused by the exchange (central office) being a long way from one or both of the modems. This problem is common in rural areas. Degradation can also be caused by slight imperfections in the cabling between the modem and the exchange [2].

Pair Gain Systems

A potentially more serious source of problems are *pair gain systems*. A pair gain system (sometimes referred to as a carrier system) is a device used by the telephone company in areas where there is a cable shortage or to improve the audio quality on long lines. Some pair gain systems are known to limit modems to

---

1. There are two main competing technologies: K56 and X2. A standard is not expected until 1998 [1].

speeds under 9,600 bps, thereby affecting the performance of fax machines as well as modems. Most telephone companies will make an effort to solve this type of problem if you can speak to the right person, particularly if it is reported as a problem with a fax machine.

It should be noted that the most popular pair gain systems in use today are digital and can actually improve the performance of modems on long lines, provided the pair gain system is installed with modems in mind. Typically a digital pair gain system will allow a V.34 modem to operate at 21.6 Kbps.

It should not be necessary to rent a high-quality telephone line to get a modem to work. PSTN modems are designed to work with ordinary telephone lines. V.34 modems are better than any of their predecessors at coping with impaired lines. They can fall back to a lower speed if they detect a momentary degradation caused for example by noise on the line and then fall forward if conditions improve. They also measure the properties of the line at the beginning of a call and adjust themselves to make the most of that particular connection.

Speed Negotiation

Modems operating to different ITU-T standards will negotiate to work at the speed of the slowest modem. For example, a V.34 modem calling a V.32 bis modem will connect at 14.4 Kbps or less. This speed negotiation is called handshaking and normally occurs at the beginning of the call. It can take between 10 and 30 seconds to complete. Twenty seconds is typical. Apart from delaying the user, handshaking occurs after the call has been answered and thus impacts on the call charge, particularly on long distance calls.

Fax Speeds

The highest standardized speed for PSTN fax at present is 14.4 Kbps. However, the majority of fax machines in use today (1997) operate at 9.6 Kbps. The ITU-T is currently working on a standard for PSTN fax operating at 28.8 Kbps, but until this standard is released, there is a big gap in speed between standard fax machines operating at 9.6 Kbps and ISDN fax machines operating at 64 Kbps.

The ITU-T standard for PSTN fax specifies two resolutions—standard and fine—corresponding to 203×98 and 203×196 dpi. At fine resolution, it takes about one minute to transmit an A4 page at 9.6 Kbps. At standard resolution, the transmission will take about half that time. The actual transmission time will depend on the density of information on the page so that an almost blank page will transmit much faster than a page of tightly packed text.

11.2.4 Data Compression

Most modern modems employ data compression to speed up data transfer (Figure 11.3). Unlike the techniques used for audio and video compression, data compression does not degrade the data quality in any way. It simply uses a shorthand technique to reduce the time spent transmitting repetitious bit patterns. A similar technique is used by file-compression software such as PKZip. The compression technique used in modems is standardized by the ITU-T and is called V.42 bis. Because it is standardized, it can operate between modems from different manufacturers.

Modem advertisements often claim a 4:1 compression ratio, giving an effective throughput of 134.4 Kbps for a V.34 modem (twice the speed of a standard ISDN connection!). What these advertisements gloss over is that 4:1 is the maximum compression ratio that V.42 bis can achieve, and it only happens when the data to be transmitted is highly compressible. In practice, compression ratios of about 2:1 are much more common when transmitting typical text and graphic files. If the data is highly random (e.g., encrypted files) or already compressed (e.g., by using PKZip before transmission), then the modem will not be able to further compress the data and there will be no increase in throughput.

Modems use on-the-fly compression (i.e., they compress the data just before it is transmitted). Sometimes it is possible to gain improvements by compressing offline before a big file transfer. For example, some files can be compressed by 8:1 or more, but V.42 bis cannot do better than 4:1 on any file. Another reason for offline compression is when you want to use encryption. Encrypting a file makes it incompressible. The solution is to compress it before you encrypt. PKZip can in fact compress and encrypt in one operation. In many instances, however, offline compression is not an option. For example, it is not an option when querying a database, surfing the Internet, or doing collaborative work with a remote colleague.

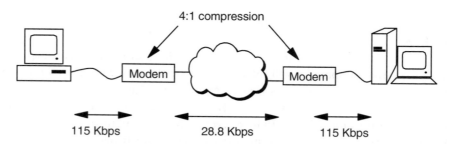

Figure 11.3 Data compression is used to speed up the effective throughput of a modem.

11.2.5 Error-Correcting Modems

All modern modems include built-in error detection and error correction. The technique used is standardized by the ITU-T as V.42 and is available on the vast majority of today's modems. It works by adding a *checksum* to each block of data sent. The checksum is a bit pattern that is calculated from the block of data. If the receiving modem calculates a different checksum for the block, it requests a retransmission.

11.2.6 Selecting a Modem

Voiceband modems vary in price from about $50 to $1,000. Even if you restrict yourself to V.34 modems, the price range is still very large: $100 to $1,000! When choosing a modem, the most important question to ask is: "What is the modem required to do?" There is quite a difference between modems connected to a corporate LAN used to accept calls from staff on the move and a modem used for occasional recreational access to the Internet. In the first case, reliability and the ability to interwork with a wide range of modem types is essential. In the second case, price will probably be the most important factor.

General Considerations

We will now look at some general considerations affecting modem choice and particular modem uses.

- *The ability to interoperate with other modems.* Top-of-the-line modems tend to be better at establishing connections to a variety of other modems [3]. If a modem is to be used regularly to call a specific service, then you will likely get away with a cheaper modem if you can find out what types of modem work well on that service. Typically modems from the same manufacturer will interwork well, as will modems utilizing the same chip set (many modems from different manufacturers use the same chips internally).

- *Ease of upgrade.* Some modems can be upgraded by downloading upgraded software into them. These modems uses *flash EPROM* memory. Modem upgrades will typically improve the performance of a modem, fix interoperability problems, or add a new feature. Other modems can only be upgraded by changing a chip in the modem, while others offer no upgrade facility at all. There are two aspects to modem upgrades: the data pump and the ancillary software. Obviously, the facility to upgrade both aspects is preferable [1].

- *Options supported.* The V.34 standard has many optional parts. In general, the more of these options a modem supports, the better it will perform.
- *Type approval.* See Section 11.8.

Modems for Teleworkers

Teleworkers will typically want a fast modem (if they cannot afford ISDN) with fax capability. Another feature that will be useful to some teleworkers is known as *digital simultaneous voice and data* (DSVD). It allows the user to speak over the connection at the same time as transmitting data, provided there is a DSVD modem at both ends. This feature, combined with appropriate software, allows collaborative working, (e.g., where two people work on the same document from two different locations discussing changes as they go).

DSVD modems create two channels from the one connection, dividing the available bandwidth between voice and data. The voice channel uses digital speech coded at about 9.6 Kbps and reduces the data transfer rate by that amount. Up to 1996, DSVD products were proprietary. The ITU-T has however ratified a collection of standards for DSVD (V.70, V.75, and V.76, August 1996).

Some DSVD modems allow a teleworker to dialup a similar modem in the office and connect to the LAN and the PBX at the same time (Figure 11.4). Security is very important here, not only for the data connection but also the voice connection to prevent intruders from making calls and having them charged to the PBX. Access can be restricted to legitimate teleworkers by the use of passwords, calling line identification, closed user groups, or a combination of these methods.

DSVD should not be confused with the slightly less powerful modem feature called "VoiceView" (by Radish Communications Systems, Inc.), which al-

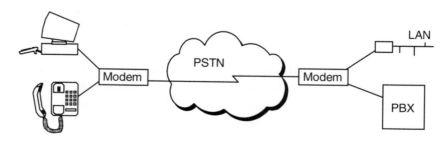

Figure 11.4 A teleworker can connect to the company LAN and PBX simultaneously using a voice modem.

lows you to *switch* between voice and data without interrupting the call but does not allow *simultaneous* voice and data.

Another facility that may be useful for the teleworker is a modem that is capable of remote management. This is a feature of some V.34 modems that allows a technical support person to configure a modem remotely, putting less of a technical burden on the teleworker. Remote configuration will only work, however, if a call can be established in the first place, making it of limited use. This type of remote configuration should not be confused with the extremely useful remote configuration options available for rack-mounted modems used in modem pools.

Modems for the Small Business

The small business will not likely be interested in DSVD or remote configuration. High on the agenda will be the ability to send fax and email messages. A modem with built-in voice-mail capabilities may also be attractive. This allows a PC and modem to be used as a voice-mail system. PC-based voice-mail systems can work out to be very cost effective because the most costly element of the system (storage) is already there in the PC's hard disk. Care must be taken because voice messages can quickly eat into hard disk space.

None of these applications require the speed of a V.34 modem unless there is a very high volume of email or file transfer. Much more important is the ability to connect to a wide range of fax machines and to connect to your email service the first time every time. A good V.32 bis modem will actually work quite well, but as time goes on there is less and less of a cost difference between V.32 bis and V.34 modems.

One-person businesses often use a single telephone line for voice, data, fax, and answering machine. There are a variety of ways of achieving this. The first thing you will require is a device that can distinguish what type of call is coming in. Such a device is often built into fax machines. In most cases this device answers the call and determines what is at the far end. If it detects voice, it rings the phone or answering machine; otherwise, it connects to the fax machine or fax-modem.

There is a conflict between the voice/fax switch just described and public voice mail. If the fax machine or voice/fax switch answers every call to determine what type of call it is, then the call can never be answered by the public voice mail (because the public voice mail is triggered by a *nonanswer*). The best solution in this case is where the phone company offers a public voice-mail facility that can store both voice and fax messages. This has the added advantage over an answering machine that people can leave voice or fax messages even while you are on the phone.

An alternative to the voice/fax switch is a feature that some phone companies offer known as *distinctive ringing*. In this case you get two or more telephone numbers with your telephone line. Each number causes your line to ring in a different way. Some modems have the ability to distinguish the type of ringing and answer the call in an appropriate way depending whether it is a fax or data call, or to leave the line ringing so that you can answer the phone in the case of a voice call [1].

Modems for Use With Mobile Phones

Analog mobile phones (AMPS, TACS, NMT) work best with so called cellular modems, as discussed in Section 9.3.2. Digital mobile phones (GSM, PCN, PCS) require a data adapter designed for the particular mobile network technology.

Modems in a Shared Modem Pool

If you need to set up a bank of modems as a shared facility (e.g., to allow a group of teleworkers or mobile staff to connect to the company LAN), you will need some serious modems to give you the reliability, manageability, and security required by this application.

Modems that can connect to a wide variety of other models and have a reputation for reliability are preferable. If there are a lot of modems involved then *rack-mounted* modems will save space (Figure 11.5). Some rack-mounted modems include multiple modems per card. In addition, it is possible to get modem chassis that connect directly to a primary-rate ISDN line. A management system, while expensive, will allow support staff to configure or reset

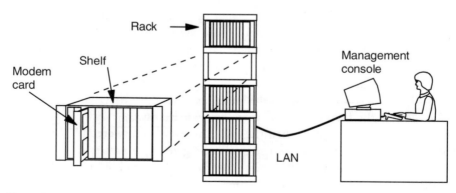

Figure 11.5 Rack-mounted modems save space in the equipment room.

the modems from a single console. In most cases, this console can be remote from the equipment room.

A number of security options are available on modems that give an extra line of defense over and above the passwords required for your computer systems. The first is password protection on the modems. The second is call-back, a feature whereby the modem requests a password and then hangs up and calls back a predefined telephone number associated with that password. Some modems allow multiple passwords and associated call-back numbers. Finally, if the telephone network provides CLID, some systems can be programmed to accept only calls from certain telephone numbers.

When using the call-back feature, it is good to arrange that the modem calls back on a different line than the one the call came in on [4]. A very simple attack on a call-back system is for the hacker to dial the system and stay on the line and mimic the PSTN tones while the modem tries to dial him or her back. The hacker thus fools the system to think that it has called him or her back and can now attempt an attack on the computer system.

Modems Used on Leased Lines

Modems used on leased lines are often significantly more expensive than dialup modems. For resilience, you can specify a leased-line modem that includes dial backup (i.e., if the leased line fails, the modem has a second port connected to a PSTN line and it automatically dials up a predefined number, which is the number of the modem at the far end of the leased line) (Figure 11.6). Yet another feature called *leased-line lookback* will automatically cause the modems to revert to leased-line operation whenever the leased line is restored to service.

It is quite common nowadays to use ISDN rather than the PSTN for leased-line backup. This is because of the greater bandwidth and the shorter call setup times. In the case of ISDN, the modem in Figure 11.6 is replaced by a suitable terminal adapter.

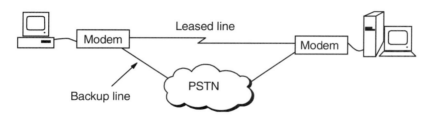

Figure 11.6 Leased-line modem with dial backup.

11.3 Multiplexers

A multiplexer is a device that allows a number of connections to share a single communication line. A very common use of a multiplexer is to allow many devices to share a leased line rather than allocating a separate line to each device (Figure 11.7). The majority of multiplexers used in private networks are proprietary equipment, meaning that the same manufacturer's equipment must be used at either end of the link.

There are two common multiplexer types: *time-division multiplexers* (TDMs) and *statistical multiplexers* (STDMs or Stat muxes).

11.3.1 Time-Division Multiplexers

Time-division multiplexers work by transmitting a few bits from each input line in turn, as shown in Figure 11.8.

The bandwidth of the composite channel must be greater than the sum of all the individual bandwidths. You will notice if you look at Figure 11.8 that a little extra bandwidth is required for synchronization bits (indicated by an "s"). The synchronization bits are required so that at the far end, the demultiplexer knows where the cycle of channels starts, thus allowing it to reconstitute the individual channels correctly.

TDM multiplexers are well suited to voice transmission because the multiplexer introduces very little delay into the transmission path and allocates a fixed (guaranteed) bandwidth to each connection. A very common voice ap-

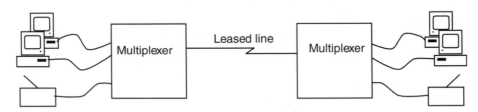

Figure 11.7 A multiplexer allows many devices share a single leased line.

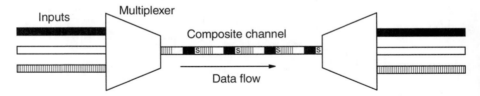

Figure 11.8 The operation of a time-division multiplexer.

plication is the interconnection of PBXs over digital lines. In this case, special low-bit-rate voice coders can be used to gain maximum usage of the private line (Figure 11.9).

In the example shown in Figure 11.9, the private line could be a 64-Kbps line, in which case the codecs would need to code the voice at 8 Kbps. Alternatively, a higher speed private line could be used along with higher quality voice coding at, say, 16 Kbps.

Video and LAN Interconnect

TDM multiplexers can also be used for the integration of video-conference connections and LAN-interconnect traffic with voice circuits (Figure 11.10). In these cases, higher speed private lines would be necessary. A high-quality video-conference requires about 384 Kbps. This is a lot of bandwidth. In cases where videoconferences are not held every day, it is very useful to be able to allocate bandwidth to the video-conferencing equipment only when it is required. The ease with which this can be done will depend on the multiplexer. Ideally, it

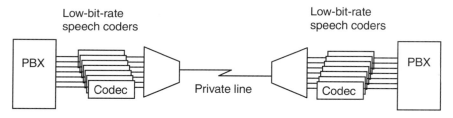

Figure 11.9 PBXs can be linked over a single private line using low-bit-rate speech coders (codecs) and multiplexers. The codecs are normally an optional addition to the multiplexer.

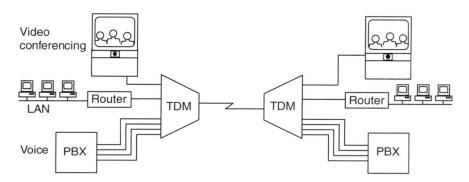

Figure 11.10 Voice, video, and LAN interconnect sharing a private line.

should not require the intervention of support staff. The private line would require at least 512 Kbps in this instance.

T1 and E1 Multiplexers

A very common type of TDM multiplexer found both in private and public telephone networks is one which combines 24 or 30 × 64 Kbps channels into a single T1 (24 channels) or E1 (30 channels) channel (see also Section 5.2.1). This style of multiplexer is called a channel bank in the United States and often includes analog to digital conversion for voice circuits.

11.3.2 Statistical Multiplexers

Statistical multiplexers are used when the data to be transmitted is bursty. They use packet-switching techniques to combine the inputs into one channel. Data is divided into short blocks and headers are added to each block (Figure 11.11). The headers allow the demultiplexer at the far end to determine which channel the block belongs to.

The bandwidth of the composite channel can be *less* than the sum of the bandwidths of the inputs because the multiplexer only allocates bandwidth when it is needed. This makes the statistical multiplexer very efficient when it comes to bursty applications, such as cash-dispenser transactions, database access, and remote monitoring equipment.

It is quite common to find a statistical multiplexer built into a TDM multiplexer, thus allowing voice and bursty data to be integrated onto a single private line. Figure 11.12 shows how this works. The output from the statistical multiplexer becomes one of the inputs to the TDM multiplexer. The rest of the TDM inputs in this example are voice channels. Multiplexers in this configuration can often be programmed to allocate extra bandwidth to data at night, when the voice bandwidth is no longer required.

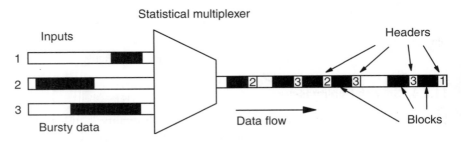

Figure 11.11 Statistical multiplexers use headers to distinguish the channels.

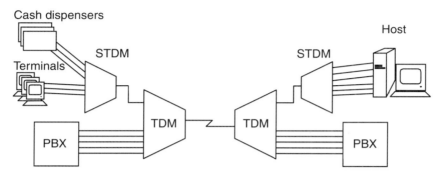

Figure 11.12 Combination of STDM and TDM integrating voice and bursty data over a private line.

The requirement for statistical multiplexers has declined over the years, as more and more applications are supported over LANs and interconnected LANs as opposed to host-terminal networks.

11.3.3 Dynamic Bandwidth Allocation to Voice

In a typical conversation between two people, each person speaks for 40% of the time. The rest of the time is spent listening or pausing. Some multiplexers make use of this fact to squeeze more voice channels out of a private line. They make use of a device called a *voice-activity detector* (VAD) to detect speech and only allocate bandwidth while one of the parties is speaking.

In theory, this technique can yield 2.5 times as many speech connections on a given private line. In practice, however, you will encounter clipping of speech if you try to squeeze too much out of your private line. Clipping refers to the loss of the beginning of a sentence because all of the bandwidth is occupied.

These multiplexers can also provide dynamic bandwidth allocation between voice and data. In a traditional TDM, you must preassign a certain amount of bandwidth to voice. If few people are using the phone, a TDM multiplexer will not be able to dynamically allocate any extra bandwidth to data.

11.3.4 Speech Quality Issues

Digitizing speech involves a tradeoff between speech quality, bandwidth, and complexity. Table 11.2 lists the characteristics of speech that combine to give an overall quality.

Table 11.2
Speech Characteristics

Characteristic	Explanation
Intelligibility	It must be possible to readily understand what the speaker is saying. This is the overriding consideration, regardless of application.
Distortion and noise	All speech codecs introduce some distortion. The acceptable level depends on the particular application.
Delay	The more complex the codec, the greater the delay. Even moderate delay can be objectionable. It often results in one person interrupting the other.
Recognizability	For general-purpose business use, it is important that the speaker can be readily recognized by the other party. This characteristic is used (almost unconsciously) as an authentication mechanism, and this aspect should not be underestimated. For internal use, speaker recognition can sometimes be sacrificed in the interests of greater cost efficiency—particularly if alternative authentication mechanisms are in place.

For a given coding technique, reducing the bit rate also reduces the speech quality. However, more modern speech coding techniques can outperform their predecessors even when using a *lower* bit rate. It is important, therefore, to judge a particular coding technique on quality measurements or a live evaluation rather than by looking at the bit rate [5].

Speech quality is usually measured in terms of overall quality. The most common unit being the *mean opinion score* (MOS)—which is an average of several peoples' opinion. Intelligibility can also be measured as a separate entity. Speech quality is highly subjective and should ideally be assessed under similar conditions to its proposed usage.

One of the most serious issues on a network that uses low-bit-rate speech is *double hop*. Consider the network shown in Figure 11.13. Communication between A and B must be decoded to analog to transit the headquarters' PBX and then re-encoded as a low-bit-rate speech signal before being transmitted to its final destination. This double encoding and decoding causes a serious degradation in speech quality and doubles the transmission delay.

The effects of double hop can be minimized by using high-quality, low-delay speech encoders operating at about 16 Kbps. (Though not all coders operating at 16 Kbps have these characteristics.) Alternatively, some multiplexer equipment allows the compressed signal to pass through a *digital* PBX unaltered, thus creating a connection that is only a single hop as far as the speech encoding-decoding cycles are concerned.

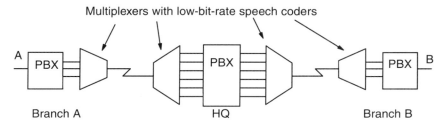

Figure 11.13 Double hop in a low-bit-rate speech network.

11.4 Network Terminating Unit

A *network terminating unit* (NTU) is a device used to terminate a digital line (i.e., a digital leased line or an access line into a data network such as a frame-relay network). In the United States, the same device is called CSU/DSU. The same device is sometimes referred to as a *baseband modem* or as a *line terminal.* The term NTU will be used for simplicity.

The main function of an NTU is to convert between the signals used on long-distance lines and the signals used on (shorter) serial cables (Figure 11.14). In addition, NTUs have facilities for monitoring the line and for fault tracing.

NTUs range in speed from 1.2 Kbps all the way up to 45 Mbps. In most cases, the speed is configurable (i.e., it can be changed). For example, many NTUs offer a range of speeds between 2.4 and 64 Kbps. Above 64 Kbps, NTUs normally come with speeds that are a multiple of 64 Kbps or multiples of 56 Kbps. It should be noted that changing the speed of an NTU cannot be done without the assistance of the telephone company because, regardless of

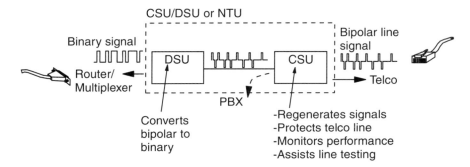

Figure 11.14 An NTU or CSU/DSU can be considered as two components as shown. In some cases, the DSU component is not needed.

whether they allow you to make the changes at your end, they will have to make changes within their network.

In addition to being used for connection to a public network, NTUs can be used as digital *line drivers* in a local network. For example, on a large campus, a pair of NTUs could be used to carry a 2-Mbps signal from one building to another over copper-pair cable. The maximum distance between the NTUs will depend on:

- The type of NTU;
- The quality of the cable (shielded is better than unshielded);
- The bit rate of the signal (the higher the bit rate, the shorter the distance).

Many NTUs can achieve distances of two or three miles at 64 Kbps over two copper pairs. At 2 Mbps, however, maximum distances can be as low as half a mile. NTUs based on HDSL technology, however, can be used at 2 Mbps over distances of two miles and more. HDSL-based products are more expensive than their non-HDSL counterparts.

11.4.1 Ownership

The rules about ownership of an NTU vary from country to country. In Europe the NTU is generally owned by the phone company, while in the United States it is owned by the person renting the service from the phone company. The advantage of the U.S. approach is that the CSU/DSU can be built into other equipment, such as routers, multiplexers, or other access devices. This saves on power supplies, cabling, and space. The advantage of the European approach is that if you order a change in your service, the telephone company is responsible for making the changes in the NTU. In some cases they can do this remotely, thus reducing provisioning times and costs.

11.4.2 T1 CSU/DSUs and E1 NTUs

The NTUs used on T1 (1.5-Mbps) or E1 (2-Mbps) leased lines differ from their lower speed counterparts because they can perform more sophisticated performance monitoring. This is possible because T1 and E1 transmission systems have a synchronization channel (time slot 0 on E1 circuits), which can be used for the measurement of the bit-error rate and passing alarm information from one end of a line to the other.

These performance-monitoring capabilities are very useful because they can indicate circuit degradation in advance of a serious fault. Faulty joints in a cable, for example, will cause an increase in the bit-error rate, but it may be several hours or days before the circuit becomes unusable. Early detection of such a fault allows the phone company to repair the joint before any loss of service is encountered.

The ability to monitor performance also has an impact on *service-level agreements* (SLAs), allowing you to check if the telephone company is keeping their side of the bargain.

Fractional Services

It is becoming common for many digital services to be provided to a business over a single T1 or E1 circuit. For example, a user might have 15 PSTN lines, a 256-Kbps frame-relay connection, and a 64-Kbps X.25 link all in the one T1 or E1 link. This can sometimes be less expensive than separate lines. In addition, it allows the possibility of performance monitoring on the link, and it allows room for expansion at short notice. (In the above configuration $15 + 4 + 1 = 20 \times 64$-Kbps channels are used. This leaves four spare channels if the circuit is T1 or 10 spares if it is E1.)

This arrangement requires equipment on the user's premises to extract the channels. This function is carried out by a multiplexer, which may be built into the CSU/DSU.

11.5 Communication Equipment for Mainframe Computers

Despite predictions about its death in the early 1990s, the mainframe computer still plays a vital role in many of today's businesses. The reasons are twofold. First, it takes considerable time and investment to rewrite software so that it can run on midrange computers. Second, there are many cases where a mainframe is the only machine with sufficient capacity and power to run particular applications.

The classical layout for a mainframe network is a star as shown in Figure 11.15. The FEP relieves the mainframe of communication-related tasks such as error control. It concentrates all the traffic coming from hundreds or even thousands of terminals. Connected to the FEP are a number of cluster controllers. A cluster controller also concentrates traffic, this time from up to 32 end-used devices such as terminals or printers.

The communication links between the FEP and the communication controllers can be leased lines, X.25, or frame-relay virtual circuits. The FEP requires special software to handle X.25 or frame relay. Many network managers

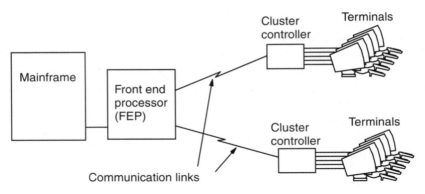

Figure 11.15 Classical mainframe network.

today are attempting to combine mainframe traffic (often called legacy traffic) with LAN traffic. This has a big economic advantage because you build one network rather than two. The big difficulty is that mainframe traffic cannot tolerate delays, while the most popular protocols used for transporting LAN traffic over the wide area (TCP/IP or IPX/SPX) have no mechanism for assigning priority to the mainframe traffic.

One solution would be to use TDM multiplexers to combine the two traffic types onto leased lines. This would fit in well if you also used the multiplexers to carry compressed voice on the same leased lines. The drawback of this approach is that TDM multiplexers, while guaranteeing bandwidth and response times to the mainframe traffic, do not allow you to share the bandwidth in an optimum way. They do not, for instance, allow the LAN traffic to take advantage of idle time in the mainframe traffic and vice versa.

The preferred solution is to access the mainframe computer over the LAN interconnect. One method of achieving this is described in Section 12.6.1.

11.6 Compression Equipment

We have already discussed data compression with respect to analog modems that use the V.42 bis compression standard when operating in *asynchronous* mode. *Synchronous* data, however, cannot use V.42 bis compression. It is quite common, for example, to find modems, ISDN terminal adapters, and CSU/DSUs that compress data when used in asynchronous mode but that do not compress when used in synchronous mode. This has implications for LAN interconnection because routers and bridges always work in synchronous mode.

Even where *synchronous data compression* (SDC) is available in a piece of equipment, it is often implemented using a proprietary technique and may not interwork with equipment from another manufacturer [6].

Synchronous data compression can be implemented in stand-alone units, or it can be integrated into other equipment, such as routers, multiplexers, FRADs, or CSU/DSUs [6]. Figure 11.16 shows a pair of stand-alone compression units at either end of a link.

When compression is integrated into other equipment, performance can suffer, particularly if the compression algorithm shares that device's main processor. Some integration schemes, however, provide very good performance by adding an auxiliary compression device with its own dedicated processor [7].

Just as with V.42 bis, synchronous data compression depends on the compressibility of the data being transmitted. Most synchronous compression equipment on the market can achieve 4:1 compression ratios, and some can go higher, provided the data is sufficiently compressible. Typical data transfers however will only yield 2:1 compression.

Limitations of Data Compression

Data compression requires an adequate processor if it is to live up to its expectations. Hence cheaper products do not tend to perform as well as their more expensive counterparts, even though they may use the same compression technique.

All compression devices have a speed limit. For many specialized compression devices, this is about 2 Mbps. More expensive equipment can, however, achieve rates of 10 Mbps and more. It is important to realize that the usual speed figure quoted is the *uncompressed* speed (i.e., the high-speed side of the device). For optimum performance, your compression equipment should be capable of at least four times the speed of your communication link.

Data compression is not designed for use with digital speech or video. For these applications, *lossy compression* is used, such as MPEG for video and the many low-bit-rate voice coders used for speech.

Data compression will not work on encrypted data. The way around this problem is to compress before you encrypt.

Data compression cannot be used on the output from a TDM multiplexer. The only way to compress in this case is to compress at the inputs to the multiplexer. This means more units of compression equipment.

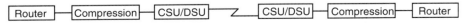

Figure 11.16 Compression equipment situated between a router and a CSU/DSU.

Compression Over Packet-/Frame-Switched Networks

Compressing data on a point-to-point link is a simple matter compared to compressing data before sending it into a packet-/frame-switched network, such as a frame-relay or an IP network [7]. In the case of a point-to-point link, you can compress everything—payload (user data), headers, and checksums. In the case of a frame-relay network or an IP network, the compression equipment must understand the protocol, ensure that the addresses in the header are untouched by compression, and recalculate checksums and other protocol information for the blocks of compressed data. This is a more complex task and requires specialized compression equipment. In these cases, it is common for the compression to be integrated into the router or FRAD being connected to the network.

11.7 Video Communication

Videoconferencing and video telephony have been growing in popularity due to the maturing of the standards, the greater availability of ISDN and other high-bandwidth services, and more recently desktop computers with sufficient power to perform the video compression in software [8].

It is useful to distinguish between videoconferencing and video telephony. Videoconferencing consists of one group of people talking to another group, while video telephony refers to one person talking to another. Full videoconferencing requires large screens, high image and audio quality, remote-control cameras and ideally an additional document camera.

It is generally accepted that one-to-one video telephony requires about 128 Kbps while videoconferencing with many people in each room requires 384 Kbps or more. To achieve these bit rates, most users employ ISDN or leased-line services. It is also possible to use frame relay, although it tends to lower the video quality somewhat and to increase the end-to-end delay in the connection.

Video Compression

All video telephony equipment employs video compression. Unlike data compression, video compression is lossy and the more you compress, the worse the image is going to appear. The compression techniques used can be based on standards such as *Motion Picture Expert Group* (MPEG) or the ITU-T standards: H.261 and H.263. Proprietary techniques also exist but generally require the same type of equipment to be used at both ends of the link. The compression standards allow for a range of picture resolutions, frames per second, and output bit rates [8]. In general, quality suffers when the bit rate is lowered.

Using compression, it is possible to achieve the same quality as a VHS video recording with a bit rate of about 1.5 Mbps (including the sound). The vast majority of videoconferencing links do not have this sort of bandwidth, so the quality will be lower than that of a video tape. On the other hand, if there is little movement the quality obtained over 384 Kbps can be very good.

Picture Resolution

Video resolution standards have their origin in the TV industry. *Common intermediate format* (CIF) is a standard resolution used on CD video and corresponds to 352 × 288 pixels (a pixel in this case is a single point of colored light on the screen). Various resolutions are derived from this as shown in Table 11.3

QCIF means quarter CIF (1/4 the number of pixels); S-QCIF stands for sub-QCIF. VGA is shown in the table for comparison. VGA is the lowest resolution commonly used on PC monitors.

11.7.2 Standards

The ITU-T has defined a group of standards for videoconferencing. The H.320 recommendation is an umbrella standard for videoconferencing over ISDN or leased line. As can be seen from Table 11.4, H.320 encompasses standards for both video and audio.

There are many optional parts to H.320 and to H.261. Equipment conforming to H.320 will interwork with other H.320 equipment, but in the worst case it may only work at 7.5 fps (frames per second) and/or with a resolution of 176 × 144 (i.e., a small image in the corner of the screen or a very blotchy image if it is displayed full screen). The H.320 standard specifies that the same frame rate and resolution must be used in both directions for any particular call [8].

The optional parts of an H.261 video encoder include pre- and postprocessing, and motion compensation[8]. These may be missing from cheaper implementations resulting in significantly poorer image quality. On the audio side, the cheapest implementations might only include 64-Kbps PCM. This is the same coding as used in the telephone network and is quite adequate for

Table 11.3
Screen Resolutions Used in Video Telephony

Format	S-QCIF	QCIF	CIF	4CIF	16CIF	VGA
Resolution	128 × 96	176 × 144	352 × 288	704 × 576	1,408 × 1,152	640 × 480

From: [8].

Table 11.4
H.320 Standards

Standard	H.320 Status	Contents
H.320	N/A	Umbrella standard containing a collection of standards for videoconferencing at bit rates of 64 Kbps to 2 Mbps
G.711	Mandatory	Audio at 64 or 56 Kbps; 3.1 kHz; the least expensive to produce (equals standard PCM as used in the PSTN)
G.722	Optional	Audio at 64/56 Kbps; 7 kHz (high quality)
G.728	Optional	Audio at 16 Kbps; 3.1 kHz (allows more room for video)
H.261	Mandatory but has many optional parts	Video compression. The mandatory quality is QCIF at 7.5 fps. The maximum is CIF at 30 fps.
T.120 series	Not part of H.320	A series of standards covering multipoint multimedia systems including desktop conferencing standards

speech. On the negative side, however, it uses up a lot of bandwidth, leaving less for the video.

Decoding a compressed video signal is a much simpler operation than encoding. For this reason the ITU-T specifications require that decoders be capable of handling higher quality signals than the encoders. This can lead to the anomaly shown in Figure 11.17, where the image seen on the low-quality product is better than the image seen on the high-quality product. It should be noted that both products comply with H.320. This anomaly is further compounded by the use of low-quality cameras on the cheaper products. When

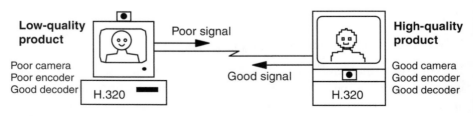

Figure 11.17 Quality anomaly in video telephony.

viewing demonstrations of video telephony products, it is very important to see how the image appears on the other end of a connection.

Video on Other Networks

Recently standards similar to H.320 have been ratified or proposed for other networks:

- H.321: ATM (or B-ISDN);
- H.322: LANs with guaranteed quality of service;
- H.323: LANs with nonguaranteed quality of service (and internets);
- H.324: PSTN with V.34 modems;
- H.324/M: Mobile phone networks (draft).

Of these standards, H.323 is a very interesting development because of its ability to work over private intranets (see Section 8.4.8) and even the Internet itself. Performance will of course be dependent on the bandwidth availability on these networks.

11.7.3 Multiparty Conferences

If more than two sites wish to partake in a videoconference you will need a multipoint bridge. Each site connects to the bridge. A multipoint bridge for videoconferencing is expensive ($40,000 for a basic model in 1997). It can be located anywhere as long as all sites can connect to it. Thus, you may lease bridging facilities from a third party.

There are two modes of operation for multiparty conferences. The simplest is where you view one site at a time. The other is where you view many sites at the one time. This mode is called *continuous presence multipoint.*

One unfortunate aspect of bridging is that the entire conference must operate at the frame rate and resolution of the poorest equipment. Thus one poor videotelephony unit can bring down the quality of an entire multiparty conference.

11.7.4 Leasing of Videoconferencing Facilities

For companies that have only occasional videoconferencing needs it can work out to be more cost effective to lease videoconferencing facilities. Typical rates are $100 to $200 per hour. Such facilities are very often to be found at universities, telephone companies, and videoconferencing equipment vendors.

11.8 Type Approval

Any equipment connected to a public network must, by law, have the appropriate approvals. These approvals aim to ensure that the equipment will not damage the network, will not interfere with other equipment, or will not pose any safety risk to the users of the equipment or the employees of the network operator. The equipment owner could very easily find themselves liable in the case of an accident involving nonapproved equipment.

The approvals are issued by a regulatory body that is normally independent of the telephone company. In many countries, this body is part of the government. Approved equipment should be clearly marked as such and should carry a certificate of compliance with national laws.

In many countries it is legal to *sell* nonapproved equipment but illegal to *connect* it to a public network. It should be noted that in many cases indirect connection of unapproved equipment to the public network is also disallowed. For example it would not be permitted to dial over the public network using an unapproved modem connected to an extension line of a PBX.

Many countries require that equipment is type approved for that particular country. Testing for type approval is expensive. In small countries, the testing costs have to be shared among a small user base, thus adding considerably to the price of the equipment. In many cases, manufacturers do not bother with type approval in small markets, thus making it difficult to build a single vendor network that spans many small countries. This fact alone leads many to use a system integrator or a one-stop-shop when building such a network.

The European Union is aiming at mutual recognition of type approvals from one country to the next. This is the case for ISDN and GSM equipment in Europe but is not yet the case for PSTN equipment.

References

[1] Navas, John, "Navas 28800-56k Modem FAQ," http://web.aimnet.com/~jnavas/modem/faq.html, 1997.

[2] Garfield, William, "Dialup Line Quality in Houston, Texas—Expecting 28800 BPS?" http://www.houston.tx.us/internet/dialup.shtml, July 1996.

[3] Kieran, Taylor, "V.34 Modems: You Get What You Pay For," *Data Communications International,* McGraw-Hill, June 1995.

[4] Goggans, Chris (STS, New York), "Information Insecurity," seminar presented in Dublin, Sept. 25, 1995.

[5] Kocen, Ross, "Voice Over Frame Relay—White Paper," ACT Networks, Inc. of Camarillo, California, http://www.acti.com/vofr.htm, 1995.

[6] "A White Paper on Synchronous Data Compression Over Wide Area Networks," Motorola Information Systems Group, http://www.mot.com/MIMS/ISG/Papers/sdc_wp.html, May 1996.

[7] Heywood, Peter, "Compression and Routers, Together at Last," *Data Communications International,* McGraw-Hill, April 1995.

[8] VTEL Corporation, "H320: A Quality Requirement Guide," http://www.vtel.com/vcnews/wpap1.html, 1995.

12

LAN Interconnect

12.1 Introduction

The interconnection of LANs is becoming more and more of an imperative for organizations with more than one site. A well-planned network of inter-connected LANs can provide the basis for all of a company's internal electronic communications and can even be used as a platform for internal voice and video communications.

12.1.1 Internetwork

A network of interconnected LANs is often called an *internetwork*. However, strictly speaking, this term should only be used when the LANs are inter-connected using routers. A typical enterprisewide internetwork includes work-stations, servers, and mainframes connected to LANs. These LANs are interconnected using communication services such as leased lines, frame relay, ISDN, SMDS, or ATM. Bridges or routers are used to ensure that data are sent towards the correct destination and that expensive bandwidth is not used to carry unnecessary traffic. Gateways will be required in some instances to convert between incompatible systems.

Figure 12.1 shows the basic requirements for interconnecting two LANs. This includes equipment with at least one LAN port and one WAN port, and a (data) communication service to provide the interconnection.

12.1.2 LAN-Interconnect Equipment

The main types of LAN-interconnect equipment are:

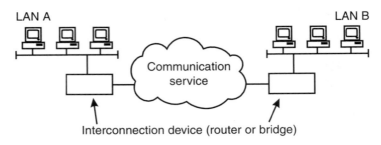

Figure 12.1 Interconnection of LANs.

- Bridges that interconnect LANs in a very simple manner;
- Routers that interconnect LANs in a sophisticated manner allowing very large networks to be built;
- Gateways that interconnect incompatible networks.

Each of these devices can be used to interconnect LANs, but they handle the packets of data differently. To be precise, they operate at different layers of the OSI model (see Section 7.4.3). A bridge operates at layer 2, a router at layer 3, and a gateway can operate at some of layers 4 to 7.

Routers are generally the device of choice for all but the smallest of networks. This is because of their ability to manage traffic efficiently and to reroute traffic if there is a failure in the network. Bridges, on the other hand, are cheaper and easier to configure. Bridges are often used on the peripheries of large networks to link small LANs or single PCs into a larger router-based internetwork.

Gateways are special-purpose devices used to overcome incompatibilities between devices or networks. They are seldom used as an interconnect device; rather, they are used as protocol converters. In the past, however, it was common to refer to routers as gateways. This has led to a dual usage of the word and some terminology, such as *default gateway* and *exterior gateway protocol* (see Section 12.5), has more to do with routers than with what is commonly referred to as a gateway.

Bridges and routers are normally dedicated boxes that can be purchased with the required LAN and WAN interfaces. These are referred to as *hardware* or *stand-alone devices*. Alternatively, they can be implemented either within a PC or a file server by adding appropriate network cards and software. This implementation is referred to as a *software router* and can provide a cost-effective means of connecting a small site into the corporate network.

In recent years most routers incorporate a bridging function in addition to the routing function and can therefore be referred to as *brouters.*

12.2 Combating the Deficiencies of WAN Links

LAN traffic, by its nature, is bursty and generally requires a fast response. This is generally not a problem within a LAN, but can be problematic over WAN links.

LANs operate at speeds of 4 to 100 Mbps, while WAN links are often restricted to 64 Kbps for economic reasons. Because the WAN link operates at a slower rate than the LAN, it is important that the interconnecting equipment prevents unnecessary traffic from passing through it. Careful use of filtering and spoofing mechanisms can greatly reduce *service advertisements,* which in turn can improve performance by leaving more bandwidth available for real traffic.

12.2.1 Spoofing

Spoofing is a very effective method of reducing unnecessary traffic across WAN links (Figure 12.2). A workstation connected to a server must respond to watchdog packets that the server sends at regular intervals. These watchdog packets are used to ensure that the connection is still active and should be kept open. If this connection is across a WAN link, then these watchdog packets will be using up valuable bandwidth or may cause an unnecessary call to be initiated across a dialup ISDN link.

A bridged link, configured for spoofing, will allow the bridge at the server end to reply to the watchdog packets as if it is the workstation, while the bridge at the workstation end will generate watchdog packets and absorb the responses. This takes place without passing any data across the link. When real

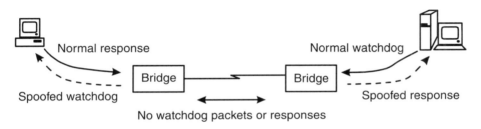

Figure 12.2 Spoofing.

data activity begins again on the link, the bridges will confirm with each other that all connections still exist.

12.2.2 Configuring Applications for the WAN

WAN links usually differ from LAN links in two respects: they generally have lower bandwidth, and they usually have higher *latency*. (Latency means delay.) Because of these two deficiencies, software applications that work well across a LAN may work very slowly across a WAN. Applications that are configured to take account of these deficiencies will perform better over a WAN.

Positioning application files on the user side of WAN links will significantly reduce the amounts of traffic passing across the links. For example, simply logging into a file server where the login command is stored on the server will generate about 10 times more packets across the WAN link than if the login command were stored locally.

If a user needs to transfer a 64-KB file from a server to a workstation and if both the server and workstation are on the same 10-Mbps Ethernet segment, the transfer will take about 100 ms. The same file being transferred between the server and workstation across a 64-Kbps WAN link would take more than 8 seconds.

Client/server applications, as noted in Chapter 4, split up applications between the server and the user's workstation (client). This has the tendency to keep data transfer across WANs to a minimum; however, a bit of fine tuning can improve things a lot.

First, bandwidth can be conserved by limiting the amount of data received by the client for any given query. For example, if a particular query results in the selection of 1,000 records, they could be fed to the client one page at a time rather than as a continuous stream of data.

Another type of fine tuning for client/server applications deals with the chatty nature of database query languages such as *structured query language* (SQL). When using SQL, a single query from the user can result in hundreds of inquiry-response transactions between the client and the server. The problem is that delays (latency) across WAN links (particularly on satellite, frame-relay, or X.25 links) get multiplied by the number or transactions. For example, 100 inquiry-response transactions across a frame-relay link with a round-trip latency of 100 ms will add 10 seconds to the response time for a single query [1].

Fine tuning of client/server applications implies software-development work. The best time to do this is during initial development. Software developers need to be made aware that their applications will be required to run over a WAN with less than ideal bandwidth and latency. If possible, software developers should have access to a testbed that simulates real-world WAN links.

An alternative way to improve the performance of SQL databases across high-latency links is to use a web server as a front end for the database server. The web server can be located on the same LAN or within the same computer as the database server, thus overcoming the latency problem. Users interact with the web server, which in turn queries the database and delivers results back to the users using web protocols. Web protocols, such as http, are designed to work well even when latency is high.

12.2.3 Choosing Between Communication Services

The type of communication link best suited to LAN interconnection will depend on the applications that will use the links. The least demanding application is email; the most demanding is videoconferencing or disk mirroring. Some applications, such as voice, mainframe access, and client/server databases, require low latency but can perform well with low bandwidth. Others, such as medical imaging, require high bandwidth but latency is not a problem.

Other issues worthy of consideration are security from hackers, ease of network reconfiguration, and the viability of integrating voice traffic over the same communication links. For most managers, the first item on the agenda will be cost. However, cost comparisons should only be made between solutions that suit the applications. For example, it would be a great mistake to choose a service based on costs alone and discover later that the network is unsuitable for the applications it was supposed to support.

Table 12.1 summarizes the WAN services used for interconnecting LANs. More information on these services can be found in Chapters 5 through 8.

12.3 Addressing

Computers communicating on a network use addresses to distinguish between each other. There are different types of address used for different purposes. Many computers end up with three or four addresses, each one used in different circumstances. Every computer on a network needs at least one unique computer or node address, plus a network address that is shared with all other computers on the same network, and sometimes a user-friendly name like "Fred" or "New_York_1," which again is unique within the network.

Table 12.1
Communication Services Applied to LAN Interconnection

Leased lines	Leased lines are among the most popular communication services for all types of private networks. They offer very low latency, high security, and in most cases as much bandwidth as you are prepared to pay for. High-quality voice integration is possible through the use of multiplexers. Leased lines are often the highest cost solution.
Frame relay	Frame relay is often significantly less expensive than leased lines but suffers from higher latency and the possibility of congestion. Voice integration is possible, but voice quality will not be quite as good as with leased lines. Network changes are usually fast and inexpensive.
SMDS	Much the same as frame relay but often offers higher speeds
Satellite links	Suffer greatly from high latency but are often significantly less expensive than leased lines, particularly in remote locations
ATM	Apart from leased lines, ATM offers the greatest bandwidth and the promise of high quality and efficient voice integration. Latency is also relatively low. In the short term, however, technology is somewhat immature.
ISDN	ISDN can be used as a pay-as-you-use service with bandwidth on demand made possible by setting up calls and combining channels on a needs basis. ISDN gives the same low latency as leased lines. Call setup delays can, however, create problems, especially when multiple channels are to be combined. The configuration of routers and bridges for ISDN requires much care and know-how to avoid high ISDN bills. ISDN is very cost effective for backup links.
X.25	Because of its low bandwidth, high latency, and high overhead, X.25 is not a strong contender for LAN interconnection. In certain countries, however, the lack of other services may force you down the X.25 road.
PSTN	The PSTN is another poor contender because of its lack of security, long call setup times, and low bandwidth. It is, however, widely available and is quite useful for remote access to the corporate network.
Public IP networks	These networks are designed for LAN interconnect. The bandwidth, security, and latency will depend on how the service provider has built the network.
The Internet	Encryption products are now available that allow a business to create a virtual private network using the Internet. Each site rents an Internet connection from their local ISP. Currently, this is the cheapest solution for an international network, but it may not be the most cost effective. Bandwidth and latency cannot be guaranteed for a start, and you may spend quite a bit of time and money ensuring that your solution is secure.

A physical address is burned into each network card by the manufacturer. Each manufacturer has a block of numbers assigned to them by a standards organization; thus, no two network cards will have the same physical address. For this reason, the physical address is often used as the *computer address*. The physical address can be referred to as the *media access control* (MAC) address. (The MAC is part of OSI layer 2, and so is the physical address.)

Data is transmitted onto a LAN with the destination address of the packet as the first part (see Figure 12.3).

Physical addresses can only be used to identify computers on a single LAN or on LANs interconnected by bridges. To enable the delivery of data across a routed internetwork it is necessary to use network addresses.

When dealing with network addresses and routers, it is necessary to define a network more strictly: A network is all of the devices, including computers, bridges, and hubs, that are connected to one port of a router. Each of the networks on an internetwork must have a unique network address for each of the protocols that it carries.

Network addresses are referred to as being logical addresses (i.e., they do not physically exist, but rather are programmed into the computers and other devices that make up the network). All computers on the one network will have the same network address. It is network addressing that makes routing possible.

We will now look at how addressing is implemented in two of the major LAN internetworking protocols.

12.3.1 Internet Packet Exchange Addressing

Internet packet exchange (IPX) is the protocol developed by Novell in the 1970s. It corresponds to layer 3, the network layer, of the OSI model. It uses an eight-digit network address and a 12-digit computer address to uniquely identify any computer on the internetwork. It appears in the format:

$$00\text{-}43\text{-}67\text{-}02 : 00\text{-}60\text{-}85\text{-}24\text{-}53\text{-}78$$

The network address is chosen by the network manager. The computer address is the same as the physical address of the network card. When a computer is initialized, it reads the physical address from the network card and reads

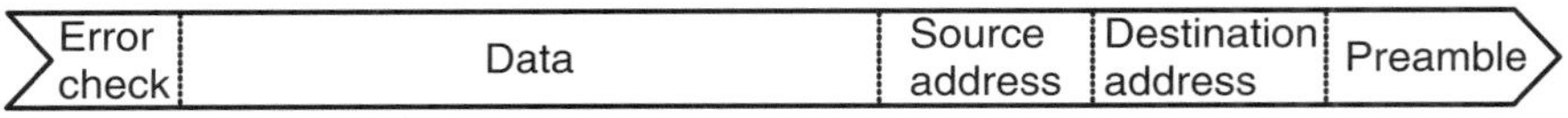

Error check	Data	Source address	Destination address	Preamble

Figure 12.3 Simplified LAN packet layout.

the network address from the network traffic. This simplifies the management of IPX addresses. The network manager needs only to configure devices that advertise the network address, such as routers and file servers.

There is no controlling body for IPX network addresses; therefore, it is the responsibility of the network manager to ensure the uniqueness of address allocation within the internetwork under his or her control. However, as there is no controlling body, there is no guarantee of worldwide uniqueness of network addresses. This may have implications if separately managed internetworks are to be merged, as can happen during company takeovers and mergers. It can also happen to companies that decide to link up disparate parts of its own network.

12.3.2 IP Addressing

IP is the Internet protocol that is part of the suite of protocols, called TCP/IP, developed by the U.S. Department of Defense in the 1970s to support the construction of worldwide internetworks. It is world famous because it is used on the Internet and has become very popular on corporate networks because of a growing trend in creating *intranets*—that is, networks with applications based on software developed for the Internet. IP corresponds to layer 3, the network layer, of the OSI model.

IP addresses are currently administered by the *network information center* (NIC), accessible at nic.ddn.mil. IP addresses are written in dotted decimal format, such as:

$$159.134.8.70$$

The one address contains both the network and computer address.

There are three classes of IP addresses defined for general use: classes A, B, and C. Class A addresses can accommodate in excess of 16 million computers; class B, 65 thousand computers; and class C, 255 computers on the network. There are two further classes of address, D and E, which are used for special interequipment messaging.

Identification of the network portion of the address is achieved by looking at the leftmost numbers in the address. Figure 12.4 shows how one address is divided into a network address and a node address depending on the class.

Subnets

IP networks can be divided into smaller units called subnetworks or subnets. The advantage of this is that a company that has been allocated one IP address

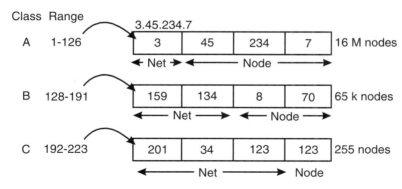

Figure 12.4 IP address classes.

need not be restricted to a single IP network. Figure 12.5 shows how a class B address of 159.134.0.0 is broken down into network, subnet, and node addresses.

Allocating Addresses

A company that has an IP address allocated to it must use only valid addresses on its computers. Although anybody can set up any IP address on a network without having one allocated to him or her, he or she cannot advertise these addresses onto the Internet. Even if a network is not connected to the Internet today, it may be in the future. It is therefore wise to get an official IP address allocation for an IP network. Alternativly, techniques such as *network address translation* (NAT) exist to translate unallocated IP addresses on a private network into valid IP addresses at the point of connection to the Internet

An IP address must be configured into each computer on the internetwork that will be using IP. This means that the network manager must assign and configure addresses for each computer. In large networks this can involve a major overhead for network administration.

Protocols such as *boot protocol* (BOOTP) and *dynamic host configuration protocol* (DHCP) enable the use of centralized servers to administer unique IP addresses to devices as required, reducing some of the address administration overheads. Computers can be set up to access a DHCP or BOOTP server on start up.

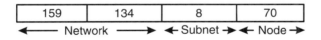

Figure 12.5 IP subnetting.

12.3.3 IP Version 6 (IPv6)

The primary drawback with IP addressing is the shortage of free addresses. They are quickly running out—there are no class A addresses remaining for allocation. Class B addresses are very hard to obtain, and class C addresses, while still abundant, are somewhat awkward to use on large internetworks—they can limit each network to 255 nodes.

The next version of IP, called IPv6, uses a 128-bit address rather than the existing 32-bit address. It is designed to support a much larger amount of addressable nodes and to allow for simpler autoconfiguration. The document RFC 1883 outlines the proposed standard for IPv6. *Request for comment* (RFC) documents are proposals published on the Internet that do as the name suggest; they set down a proposed outline and request a response. Many of the Internet standards have evolved from RFCs.

12.3.4 Domain Name System and Network Information System

Names like *ibm.com* and *nic.ddn.mil* are registered names used on the Internet to identify computers. When a computer user issues a command such as *ftp ibm.com,* an association must be made between ibm.com and its actual IP address. This association can be written in a text file, usually called HOSTS, on the user's computer. It is much more efficient, however, to install servers with searchable databases containing these name-address associations. Two common methods are *domain name system* (DNS) and Sun's *Network Information System* (NIS). IP, operating on the computer that issues the command to access ibm.com, sends a message to a DNS server requesting the associated IP address.

12.4 Bridges

A bridge is a LAN-interconnect device used for interconnecting similar LANs. Bridges can be classified as either local or remote. A local bridge provides a direct connection between multiple LAN segments in the one building or campus area, while a remote bridge connects LAN segments over a long distance using communication links.

Bridges are *protocol independent;* that is, the decision to forward a packet is made without regard to the upper-layer protocols being carried within the packet. Bridges fall into different categories, as there are different requirements from a bridge depending on the types of LANs it is interconnecting.

12.4.1 Transparent Bridges

Bridges used in Ethernet networks are transparent to the computers using them. That is, the computers on the bridged segments are not aware of the bridge or that there is more than one LAN segment. A transparent bridge builds up a table of computer addresses, as it reads them, on each of its LAN ports. In this way the bridge learns the addresses of computers on each of its ports. Figure 12.6 shows a transparent bridge and a set of tables related to its ports.

The bridge will forward those packets containing destination addresses that it knows exist on another port or a destination address that it has not yet learned. All other packets are blocked from crossing the bridge because they do not belong on the far side. In Figure 12.7(a), a packet from computer B to computer A is forwarded across the bridge. In Figure 12.7(b), the packet from computer B to computer T is blocked by the bridge as both computers B and T are on the same port.

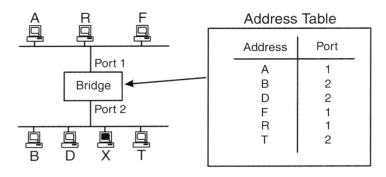

Figure 12.6 Transparent local bridge configuration.

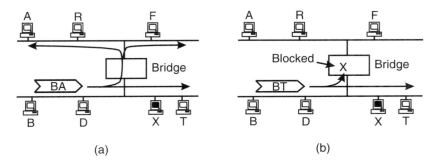

Figure 12.7 Packet (a) forwarding and (b) blocking.

Entries in the tables are erased after a period of time if the address relating to the entry has not been active within that time period. This is referred to as *aging*. The aging time can be set by the network manager but typically defaults to 15 minutes. Aging is necessary so that the bridge can keep up to date if equipment is moved from one segment to another or is removed. The downside of aging is that it increases the traffic on (expensive) WAN links. This is because packets with destinations on the same side of the bridge are allowed across the bridge until the bridge has relearned that address.

12.4.2 Transit Bridges

These connect two LAN segments of the same media type (e.g., Ethernet to Ethernet) via a different media type (e.g., FDDI). The source or destination addresses of the packet cannot be on the bridging media itself (i.e., the FDDI network in this case). This type of bridging can be used to enable an FDDI ring to act as a backbone, interconnecting several Ethernet networks.

12.4.3 Source Route Bridges

Source route bridging (SRB) is used in token-ring networks. Rather than the bridge building up tables of the location of computers, the computers send out route-determination packets as broadcasts to identify the optimal path to a required destination. The broadcast is propagated, by the bridges, onto every interconnected ring searching for the destination address. Information including bridge and ring numbers is written into the packet as it passes. On receiving the packets, the destination computer sends each packet back along the same route that it came.

The first packet to arrive back at the originating station is deemed to have traveled the optimal path. The routing information gathered on its way is now extracted and used for routing all subsequent packets for that destination.

With SRB, it is the workstation that makes the routing decisions, so the bridges require very little functionality. Consequently, they are inclined to have a higher throughput and cost less than transparent bridges.

12.4.4 Bridging Ethernet and Token Ring

It is often necessary to interconnect Ethernet and token-ring LANs. Although using a router is the preferable method, a number of bridges are available to achieve this.

This type of bridge must take into account differences between the packet-handling methods of these two media. Some of the differences are:

- Ethernet reads address bits in a packet in the opposite way to token ring. That is, an address of 1234 in Ethernet will be read as 4321 in token ring.
- The maximum packet size for token ring is almost three times that of Ethernet.
- Token-ring packets carry priority information, which Ethernet packets do not carry.

The two bridging options are:

- *Translational bridges:* These translate between token-ring and Ethernet formats by reordering the address bits, setting a maximum packet size and removing priority information from each packet as required. This is not a perfect method and can result in packets being discarded by the bridge.
- *Source-route transparent bridges:* SRT bridging implements both the source route and transparent bridge algorithms. IBM introduced the concept of SRT bridging to attempt to overcome some of the deficiencies of the translational bridges. To introduce SRTs in a network, existing SRBs may need hardware or software upgrades.

12.4.5 Bridges and Network Resilience

A potential problem arises when three or more LANs must be interconnected and network resilience is required so as to minimize the effects of link failures. Figure 12.8 shows a network of three Ethernet LANs interconnected using three remote bridges.

In this configuration, a loop has been formed between each of the LANs. The 802.3 specification for Ethernet does not allow loops in a bridged internetwork, as they will greatly degrade performance by allowing broadcast-packets to be continually transmitted around the loop. To overcome this problem, while still allowing for the insertion of redundant paths between LANs, a protocol called the *spanning tree algorithm* is used.

The spanning tree algorithm identifies redundant paths between LAN segments and blocks them, thereby removing possible loops. The bridges do

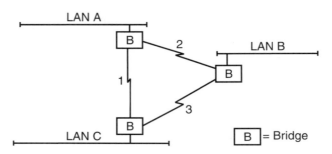

Figure 12.8 LAN interconnection with a redundant link.

this, automatically, based on user-configurable *path cost, bridge identifier,* and *port identifier* parameters.

Each bridge will block ports on paths that have been identified as being redundant. The blocked ports remain active and receive packets but do not forward them. If a link in the network fails, a message will be sent out by the bridge (or bridges) that detect the failure. The algorithm will be rerun automatically to attempt to re-establish the entire network by unblocking some of the redundant routes.

In Figure 12.8, link 3 might be the link that is selected to be blocked. This WAN link and the two associated bridge ports will be inactive for long periods. This can represent a significant cost if the link is a dedicated leased circuit. It is much more cost effective to use a dialup PSTN or ISDN connection as the redundant circuit.

Some bridge vendors have developed alternative means for using the redundant links identified by the spanning tree algorithm while not breaking the 802.3 rules. Vitalink, for example, developed what it calls *distributed load sharing* (DLS). This allows the network manager to specify a link as a DLS link. This link can be used by the bridges to forward traffic that is destined to go only between the two LANs connected by that bridge.

12.5 Routers [2]

Routers are used to interconnect networks of similar or different media types. Routers operate at layer 3 of the OSI model. They read the source and destination *network* addresses contained within the packet and determine the best route to send it. Figure 12.9 shows a possible routered internetwork interconnecting FDDI, Ethernet, and token-ring networks.

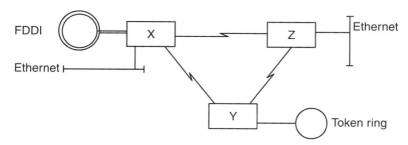

Figure 12.9 A routered internetwork.

The router maintains a set of routing tables, which are lists of routes that it has *learned* are available via its different ports. Figure 12.10 shows some simplified tables.

12.5.1 Routing Protocols

Routing protocols are those protocols that are used between routers to keep their routing tables updated and maintain the integrity of the network. Examples of these are:

- *Routing information protocol* (RIP);
- *Open shortest path first* (OSPF);
- *Interior gateway routing protocol* (IGRP);
- *Exterior gateway protocol* (EGP);
- *Border gateway protocol* (BGP).

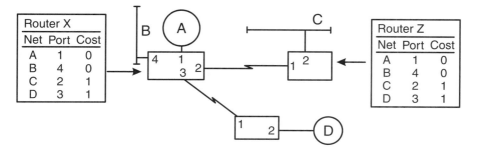

Figure 12.10 A routered network showing routing tables.

Note: The protocols used in dealing with routers fall into two categories: *routed* and *routing* protocols. Routed protocols are protocols such as IP, IPX, and AppleTalk, which operate between the two end stations and are routed over an internetwork. Routing protocols, on the other hand, are not the *means* of transmitting data. They are rules about what sort of information routers should share and when to send this information.

Unlike the transparent bridge, which builds its tables from the data packets that it reads from the network, routers, using routing protocols, advertise the networks accessible to them to all other routers on the internetwork. Consequently, routers learn about the internetwork and adapt to changes in it, such as the loss of a WAN link, much more quickly than bridges.

Routers do not age out their tables, but rather make changes to the tables dynamically, to reflect changes in the network, such as a new router or link coming on line or an existing router or link going down. A network manager can also write static routes into a router's routing tables. These are routes that the router may not be able to learn because some routers on the internetwork may be using a different routing protocol than others.

In a mesh network, a router will find that it has a number of possible routes to any particular network. The router will use a cost factor to determine which of these routes to use. The simpler routing protocols, such as RIP, select the route with the lowest hop count (i.e., the number of routers to be crossed). More complex routing protocols, such as OSPF, will use a number of user-definable parameters, such as speed and usage of the links, in determining the cost of using a particular link.

The algorithms used in routing protocols can be broken into two types: distance vector and link state.

Distance Vector Algorithms (RIP and IGRP)

Distance-vector algorithms keep the best route to any destination in their tables. When new information arrives to suggest a better route to a destination, the previous entry will be overwritten. The best route is calculated based on the shortest distance to the destination network.

RIP calculates the shortest distance based on the hop count; that is, the number of routers to be crossed to reach the destination. A hop count of 16 is deemed to indicate an unreachable network. This has a limiting effect on the size of an internetwork.

Distance-vector routers send out update messages, consisting of their entire routing tables to each of their neighbors. In large internetworks, such advertisements can use up a lot of bandwidth.

IGRP uses bandwidth, delay, and load in calculating the best route. It sends its broadcasts every 90 seconds.

Link-State Algorithms (OSPF)

Link-state algorithms build up a picture of the entire network and use a range of metrics to calculate the best route based on cost to a particular destination for each packet. The so called link-state advertisements only contain information about network *changes* and are sent out when needed rather than at set intervals. Although they are more complex and require more memory in routers, they are popular because of the greatly reduced advertisement traffic on large internetworks. They can support much larger internetworks than RIP.

Exterior Routing Protocols (EGP and BGP)

An *autonomous system* is the term used to describe an internetwork of routers that use a common *interior gateway protocol* (IGP) and is under the control of a single administrative entity. RIP, IGRP, and OSPF are all IGPs.

Exterior routing protocols are used in interconnecting two or more autonomous systems. Figure 12.11 shows two autonomous systems interconnected using an exterior routing protocol.

12.5.2 Default Gateways

A router must know a route to the destination network, or it will discard packets with that network's address. The only exception to this is where a router is nominated as a default gateway. A default gateway is a router that connects this network onto another internetwork, such as another section of the company network in another country or the Internet. Consider the router network shown in Figure 12.12.

Routers W, X, Y, and Z advertise their respective networks to each other. Router V is also involved in these advertisements but does not advertise the (vast) multitude of networks that it is aware of on port 1. W, X, Y, and Z are configured to use V as the default gateway, so they will direct to V all packets for networks unknown to them. V will then either forward or discard these packets.

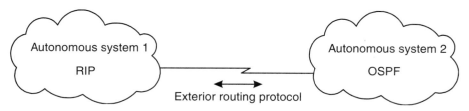

Figure 12.11 Using an exterior routing protocol.

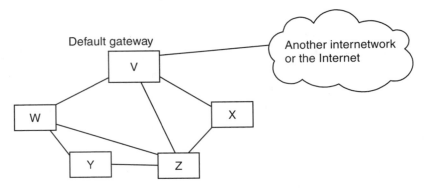

Figure 12.12 Router network with default gateway.

One drawback of this approach is that router V may have a very heavy workload. There are, however, compelling advantages.

1. The routers W, X, Y, and Z do not have to maintain very large routing tables. This reduces the amount of memory required for these routers and increases the packet throughput rate as there are less entries to search.
2. The routing advertisements between routers is simplified, reducing the amount of bandwidth being used by advertisements.
3. Router V can be used to implement security between this WAN and the outside world by filtering incoming or outgoing traffic based on the source or destination network addresses, the source or destination computer addresses, or the protocol being carried by the packets (firewall function).
4. Routers W, X, Y, and Z can be lower specification routers than router V, which can greatly reduce the cost of the entire router setup.
5. Complex network management is centralized on router V rather than across all of the routers.

12.5.3 Brouters

Brouters combine the functionality of both bridges and routers into one box. A brouter will check to see if it can route the protocol being used by the packet under consideration. If it can't, then it will attempt to bridge it. Protocols such as Microsoft's NETBUI (as used in Windows for Workgroups and Windows

95 networks), IBM's SNA, or Digital's LAT are not routable. Networks that support a combination of routable and unroutable protocols need to use brouters. Many hardware-based routers, such as those made by 3Com, Cisco, and Wellfleet, support bridging of nonroutable protocols and therefore are brouters. Some software-based routers, such as Novell's Multiprotocol router, do not however support bridging along with routing.

12.5.4 To Bridge or To Route?

While a gateway has a very definite purpose on an internetwork, the decision of whether to use bridges, routers, or a combination of both can be a difficult one. Most people involved in the area would agree that bridges and routers have moved away from the clear distinction between the two technologies, and a convergence is taking place to the single boxed bridge/router (brouter). Distinctions based on performance or throughput have all but disappeared. There are, however, specific instances where bridging is preferred to routing and vice versa or where there is no choice in the matter [2].

Bridging Advantages

Bridges hold some advantages over routers.

- Bridges are much simpler to configure than routers. A router requires, at the very least, each port to be configured. Basic transparent bridges are, literally, plug and play—remove it from the box, connect it to the network, and power it up.

- Bridges are protocol independent while routers are protocol dependent. Hence, bridges can forward nonroutable protocols such as Digital's LAT.

Router Advantages

Routers, however, can choose the best path that exists between the source and destination addresses; transparent bridges cannot. In addition:

- Routers can reconfigure after changes in the internetwork much more quickly than bridges, resulting in reduced service loss.

- A routered internetwork can support an unlimited number of computers, while a bridged internetwork is constrained to within several thousand computers.

- Routers provide a barrier against *broadcast storms.* A broadcast storm can disable an entire bridged internetwork. Transparent bridges *must* forward broadcast traffic. Routers block broadcasts by default.

- Routers act as a buffer for service advertisements from such equipment as file servers and print servers. If there are several servers on one network (i.e., on one router port), they will broadcast their service advertisements, each at their own time period. The router collects all of the information and sends it out on its ports in one block. A bridge will send the advertisements as it receives them. The router, therefore, makes more efficient use of network bandwidth.

- Routers support packet fragmentation and reassembly. If a router is required to forward a packet onto a network that has a maximum packet size smaller than the packet being sent, then the router will break the packet down and send the separate fragments. The receiving router will reassemble the fragments into the original packet. A transparent bridge will drop a packet if it is too large for the connected network.

- Routers can provide feedback to other routers and computers about traffic congestion or loss of packets; transparent bridges cannot.

Integrated Solutions

A network manager might use bridges in remote sites because of their ease of implementation and low traffic requirements, while routers might be used to provide a reliable, self-healing backbone, as well as a barrier against inadvertent broadcast storms in the local networks.

If you have a very large, meshed local internetwork (for instance, a campus consisting of several buildings and running routable protocols), routers provide superior segmentation and more efficient traffic handling than bridges.

12.6 Gateways

Gateways operate at layers 4 to 7 of the OSI model and carry out protocol conversion between connected networks or applications. They are not LAN-interconnect devices as such and thus they seldom include a WAN port. Gateways tend to be used for one specific protocol conversion (as opposed to being able to convert any protocol into any other). They often consist of software installed on a general-purpose computer.

Next we describe two types of gateways: an SNA gateway and an email gateway. The SNA gateway is one particular instance of a gateway used for communication between incompatible networks.

12.6.1 SNA Gateway

SNA is a networking protocol developed by IBM for its mainframe networks. SNA gateways are used to allow PCs or workstations to communicate with IBM mainframe computers over an internetwork. The workstation runs a 3270-terminal-emulation package such as "IrmaWIN" to act as a terminal off the distant mainframe via the gateway (Figure 12.13). The gateway is attached to a FEP via a token-ring network. The FEP is channel-attached to the mainframe computer.

Data from the workstation is sent via IPX onto the LAN. The gateway receives the IPX packets, and software running on the gateway carries out the conversion to SNA and forwards the data onto the mainframe via the FEP on the token ring. A PC with a network card and suitable software can be used as the SNA gateway.

The token-ring network may only be a single *media access unit* (MAU) cabled to the gateway and FEP, or it can be a larger ring accommodating a number of gateways.

12.6.2 Email Gateway

An email gateway converts email messages from one mail format to another. For example, a company might use a proprietary email system such as cc:Mail or MS Mail internally. To communicate over the Internet, it will need a gateway to convert their messages to SMTP (the email protocol used on the Internet).

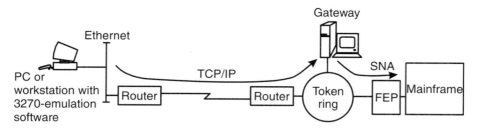

Figure 12.13 An SNA gateway supporting 3270-emulation to an IBM mainframe.

In Figure 12.14 the user creates a message using cc:Mail and sends it to the email gateway. The email gateway will receive the message and forward it in SMTP format onto the Internet via a router. The receiving network may have a gateway to convert SMTP to another proprietary mail system, such as MS Mail.

12.7 Network-Management Systems

A network-management system consists of managed devices such as hubs, bridges, and routers that can send alerts, called *trap messages,* when certain events occur, and a management station to collect these messages and to act on them. Trap messages are issued when one or more user-defined thresholds are exceeded. Events such as traffic levels on a link or CPU utilization in a router can be monitored and set to trigger a trap message. The devices running the network-management software are called *agents.* Agents gather information about the devices in which they reside and store this information in a management database.

The management station can be set to react to traps from agents by notifying an operator, simply logging the event, or initiating an automatic system repair. Management stations can also poll agents to check the values of certain variables to generate statistics. Polls can be issued manually or automatically.

An effective network-management system should not introduce so much agent traffic so as to reduce the efficiency of the network that it is managing, yet it should be aware of critical events taking place on the network. A balance is required, therefore, between the polling rates of the management station and the threshold levels for agents to send trap messages.

The collecting of network-management information is controlled by a network-management protocol such as simple network management protocol (SNMP).

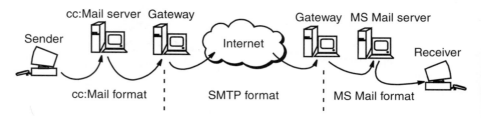

Figure 12.14 Email gateways.

12.7.1 SNMP

SNMP is protocol independent, and therefore it can be used over IP, IPX, Appletalk, and many other transport protocols. Many devices, hubs, bridges, and routers, are sold as being SNMP compliant or have SNMP management adapters as an optional extra. This means that they have a section of software or firmware that knows about this device and can act as an SNMP agent within it. The basis for information held for any device is the *management-information database* (MIB).

12.7.2 MIB-I and MIB-II

MIB-I, released in 1988, defines 114 management objects for a device, while MIB-II, released in 1990, defines 171 [2]. Both MIB versions use the same naming conventions, so devices of both types can interwork on the one internetwork. However, a MIB-I management station will not get the full range of possible information from a MIB-II device and vice versa.

References

[1] Jessup, Toby, "WAN design with client-server in mind," *Data Communications International,* August 1996.

[2] Cisco, "Univer CD—Online library of product information," Vol. 2, release 11, 1996.

13

A Business Approach to Communication

13.1 Introduction

In this chapter we begin by explaining the strategic role that communications technology plays in today's business. We then look at the ways a company can analyze its business, assess current technology, and produce a plan for its communication systems that will enhance the operation of the business.

A distinction must be made between applications and the underlying communication systems that support those applications. For example, email is an application that depends on an underlying communication network. Applications must match business needs; technology must support these applications (and future applications where possible) in an efficient manner. This chapter is mainly concerned with choosing suitable applications, while Chapter 14 deals with the underlying technology.

13.1.1 The Expanding Role of Communications Technology

The role of communications technology in underpinning a wide range of business activities is often underestimated. Yet it only takes something as simple as the failure of some PSTN lines, or maybe even a fax machine, to bring communication into sharp focus.

As technology continues to develop, the role of communication is expanding well beyond the provision of basic services. Now even the smallest business can take advantage of advanced services such as automated voice response systems, computer telephony integration, and the Internet. For larger companies, the choices are greater still, particularly as multimedia applications continue to develop.

13.1.2 Communication Is a Business Issue

Information and communication technologies (ICT) have a role to play in practically every area that influences a business' competitive position. As the range of options expands, and their cost-effectiveness increases, businesses' dependence on such systems continues to increase. Indeed, the full range of possible applications is far greater than many businesses recognize. The net result is that the optimum deployment and use of information systems is now as much a business issue as a technical one.

A Business-Focused Approach

While the need for strategic business planning has been recognized for many years, the pivotal role of information systems, as an integral part of this process, has been recognized comparatively recently. It is only more recently still that the need to consider communications technology as a distinct entity has been recognized.

Considering communication in the context of business planning involves a paradigm shift in how communication services, and information systems in general, are viewed. Traditionally they have been treated as support functions that, at best, need to be aligned with predefined business objectives. This has often meant that communication projects were inwardly focused and supportive of existing structures rather than being focused on improving the way business is done. A business-focused approach ensures that the full potential of communication systems can be critically evaluated in an overall business context. It recognizes that decisions regarding their deployment are ultimately business decisions and must be justified using standard commercial criteria—often by writing a formal business case.

Communications technology cannot be regarded as a panacea for existing organizational or functional deficiencies. It is best deployed as part of a comprehensive strategic review process, where all relevant aspects of the business are critically examined. This avoids the danger of automating, or otherwise facilitating, existing business functions or processes, which may not be optimized to meet business needs. It should also ensure maximum return on the investment required and, critically, ensure sufficient flexibility to meet evolving needs.

Failure to recognize the strategic role of communications technology, within the broader context of a developing business strategy, can mean that communication systems are assigned a low priority or implemented within a totally in appropriate framework. Worse still, they may be allowed to evolve in an ad hoc manner with no overall strategy or control mechanisms. While the more obvious communication needs may be addressed, maximum value will rarely be added to the business. Neither is there a guarantee that the resultant communication solutions will be the most efficient or cost effective available. In addition, the ongoing evolution, necessary in a changing business and technological environment, may not be adequately provided for. This can mean that opportunities go unrecognized, threats uncountered, and evolutionary concerns unaddressed.

Barriers to a Business-Focused Approach

Many businesses will be slow to recognize, let alone implement, the changes in managerial and organizational culture required for a business-focused approach to be successfully implemented. It will almost always require a strong commitment from senior management if it is to be successfully implemented. For example, technical managers will be reluctant to cede control of "their" territory to "bean-counters" and other miscellaneous riffraff. Business managers, meanwhile, may be loathe to acknowledge any role for "techies" in developing strategic initiatives. Such organizational politics are possibly the greatest impediment to full integration of the communication function with other inputs to the business strategy process.

13.1.3   A Communication Strategy

If new technologies and applications are to be fully exploited, a well-thought-out strategy is required. A particular challenge is to ensure that the business is aware of new developments in the communication area and can identify resultant applications from which real tangible business benefits can be gained. No business is too large or too small to benefit from such a strategic approach. Neither does the particular process need to be complex or time consuming.

Objectives of a Strategy

A key objective of such a strategy should be to ensure that communication adds maximum value to the business. This in turn means that the enabling role of communications technology, as discussed in Section 13.2, must be fully understood. This is particularly important with larger organizations where the range of options might be extensive.

A second objective should be to achieve adequate communication solutions in as short a time frame as possible. Business planning cycles have decreased dramatically during the 1990s as businesses strive to adapt to market changes more rapidly. This means that the time available for implementing communication solutions has also shrunk. An extensively planned, optimum communication solution is of little use if, for example, the time taken to prepare and implement it means that a market opportunity is lost. This topic is discussed further in Section 13.6.1.

Advantages of a Communication Strategy

A business communication strategy provides a framework within which technology choices and decisions can be made. A well-considered communication strategy ensures that:

- *Business requirements drive communication solutions.* Communication solutions should address clearly identified business requirements or further primary business objectives.

- *The enabling capabilities of communication systems are fully considered.* For example, they can increase internal efficiency, improve external communication, and even facilitate new organizational models.

- *Changing business needs are responded to appropriately.* Having identified opportunities/threats, it is important to be able to respond to them quickly and appropriately.

- *Proposals are assessed and prioritized using business criteria.* Any investment in communication systems should be considered solely in terms of added value to the business. This implies that a business case (Section 13.6) should always be written.

- *The most cost-efficient solutions are chosen.* There are often a range of alternative solutions to a business communication requirement. Many differences are not solely technical. Examples include alternative service providers, private networks versus VPN solutions, and leasing versus buying.

- *Changing business, regulatory, and technological environments are continually tracked.* A company that doesn't have structures in place to identify and act upon ongoing communication opportunities and threats risks losing market share to more communication-aware competitors.

- *Technology is deployed optimally.* This can only be achieved in light of clearly identified business objectives and formalized business plans. There can be a tendency for technical specialists to recommend the most technically elegant solution to a particular communication requirement. In many cases, a more pragmatic approach might result in considerable cost savings.

- *Sufficient flexibility is built in to cope with changes.* Change can be either planned or unexpected (e.g., mergers, new partners, new markets, or new locations). Flexibility is required to ensure that the business is not unwittingly locked in (either on a contractual or technological basis) to particular working methods, particular suppliers, or service providers. It should also help to avoid early obsolescence and facilitate upsizing, downsizing, or moving.

13.1.4 Communication as Part of a Strategic Review Process

Communications technology is are best deployed as part of a comprehensive strategic review process. The primary purpose of such a review is to get the business to work more effectively, efficiently, and flexibly. For example, this might mean bringing products to market faster, reducing costs, or facilitating a quicker response to changing customer requirements.

One element in this process is to fully consider the range of options made possible by ITC in general. Another is to identify any restrictions that might have a material effect on any proposed solution.

At the time of writing, there is considerable interest in process-based methodologies such as total quality management, business process reengineering, and continuous improvement. Particular key business processes are selected and then analyzed to identify desired improvements, how they might be achieved, and the cost of so doing. For example, a process might be the handling of an order by the sales department. Part of the strategic analysis would then be to consider how this particular process could be improved through the optimum deployment of ICT.

From a communication-services perspective, there does appear to be some advantages with process-based methodologies, particularly because the scope of any particular proposal need not be restricted by the existing organizational structure or working procedures. However, no endorsement of any particular technique, or indeed the overall process approach, is intended. The overall business strategy area is a developing, and sometimes controversial, science and, hence, the recommended methodologies and even the terminologies are continually evolving. A more detailed discussion of this topic is beyond the scope of this book and the reader is referred to [1–6], which give a chronological insight into this evolving area.

13.2 Enabling Role of Communications Technology

Over the last three decades, communications technology has brought about fundamental changes in the manner in which business is conducted. There has been an enormous increase in telesales and telemarketing. Electronic ordering and payment for goods is beginning to percolate down to smaller businesses, increasing their efficiency and responsiveness. Businesses that provide information services, such as electronic magazines, online catalogues, databases, and encyclopedias, depend on communication for almost all of their business activities, including advertising, sales, information gathering, and distribution of the service. By means of the Internet or fax-back systems, general queries can be answered 24 hours a day, seven days a week. Integrated voice-response systems can quickly guide telephone callers to their correct destination without any human intervention.

13.2.1 Time, Place, and Language Independence

An important aspect of today's developments in communication is the degree of freedom they give people in terms of time and place. For example, the telephone gives freedom in terms of place; the mobile phone increases this freedom. Email, voice mail, and fax give freedom in terms of time, making it easier to catch people and easier to communicate with people in different time zones.

Language independence can be addressed today through the use of email and translation software. The same is also true of fax if the print quality is sufficiently good. The results are far from perfect but will no doubt improve with time. Simultaneous voice translation is also being researched. Regardless of the development of translation software, many nonnative speakers prefer to communicate by email or fax rather than by telephone because it gives them the thinking time they need for translation.

These are all important factors in the search for new markets, lower cost labor markets, and new suppliers and distributors.

13.2.2 New Workforce Profile

ICT can provide the opportunity to radically change the profile of a business' workforce. Such changes do not occur overnight, and they are more often found in new businesses than in existing ones. The two main trends are the reduction in the size of the required work force (brought about by efficiencies created through information systems) and the increase in teleworking. Improved communication also makes it easier to contract work out to other businesses with the assurance that they are as contactable as a person working beside you.

Virtual Organizations

Communications technology is changing the very structure of some companies. With technologies such as videoconferencing and multimedia, it is now possible to have virtual organizations comprised of individuals or small groups of people, geographically dispersed yet linked seamlessly together.

Call Centers

A call center is a place where a number of people are employed to answer sales or customer-support calls. A large number of organizations have established call centers as their primary customer interface. Call centers allow entry into foreign markets without the need to establish a physical presence in the foreign country. They allow a business to select its country of operation on the basis of prevailing wage rates, tax incentives, and telecommunication facilities and tariffs. For large organizations, there exists the possibility of establishing call centers in different time zones so as to provide a 24-hour service without the employees having to work unsociable hours.

Teleworking

A growing number of businesses employ people who telework from home and perhaps visit the office once a week. Advantages exist for the employer, who saves on accommodation and other overhead costs, and for the employee, who saves on traveling time and who sees more of the family. Advocates of teleworking claim that the teleworker is actually more productive than an office worker.

Teams of employees working from dispersed locations can be established without the need to change locations to work on a project. The primary means of communication can be electronic. In this way, much time can be saved in getting the project off the ground. Journalists using laptops and modems rarely need to call to the office in person.

13.2.3 New Working Methods

There are a number of ways in which communications technology has changed working methods.

Telesales

The door-to-door salesman has virtually disappeared and has been replaced by telesales personnel. The telesales staff themselves will see further changes as their telephone systems are integrated with their computer systems.

Electronic Mail Order

Paper-based mail-order catalogs are on the verge of decline as they are converted to electronic format on the Internet.

Electronic Distribution

Newspaper publishers can establish small publishing facilities in a number of key locations. The main body of the paper is distributed electronically to these locations, while news of local interest can be added locally. This enables publishers to get their papers on the streets as quickly as possible and save on long haul transportation costs while customizing their product for each market.

Outsourcing

There is considerable interest in business today in outsourcing noncore business activities. Communication systems increase the number of functions and processes that a business can outsource. For example:

- Services to employees—wages, health insurance;
- Services to customers—help desks;
- Sales services—call centers.

13.3 Identifying Options To Maximize Business Advantage

The identification of options requires a broad knowledge and understanding of current technologies, a good knowledge of the business, and plenty of creativity.

Whenever an attractive option emerges, its technical feasibility should be checked by a specialist, its financial viability must be checked, and it must be achievable within the constraints imposed by the structure of the organization (Figure 13.1).

13.3.1 Forming a Team

The foregoing approach requires specialized knowledge of information technology, communication capabilities, individual business units, and the particular

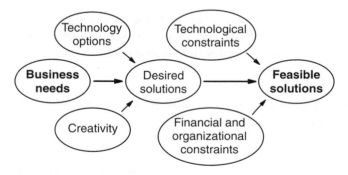

Figure 13.1 Planning solutions to business needs.

strategic methodology in use. It is rare that any one individual would have the necessary breadth of knowledge and experience to successfully complete this task. Therefore, a team approach, with specialists from a number of disciplines, is recommended. The team approach also limits the risk of strong-vested interests skewing the assessment process and possibly precluding particular options from consideration.

End users of existing systems can be a very useful source of ideas on how things can be improved and can add a much-needed touch of reality to discussions. Look out for those who make an effort to get on top of their job.

A small business that does not have all of the expertise inhouse might use a consultant to help design its communication facilities. While the consultant may be knowledgeable about the business sector in general, he or she may have limited knowledge of the particular business in question. It is important, therefore, that a consultant works as part of a team with the relevant business managers.

Problems can arise if a supplier is used as a technical expert on your team. The supplier will gain valuable inside information about your financial situation and, worse still, will get to know your innovative ideas. This latter point is important because innovative ideas give maximum return on investment while you are the only one implementing them. A supplier who sees the worth in your ideas and the potential sales they can generate for his or her company will not be slow to suggest them to your competitors. A supplier may also lead you astray in his or her recommendations if it will lead to further sales of his or her equipment.

13.3.2 Studying the Business

It is fundamental that any communication plan is based on the needs of the business. One starting point might be a critical assessment of current working procedures and the various interactions with customers, suppliers, and employees. Each area should be evaluated and an assessment made as to their overall effectiveness, particularly in light of the stated business goals. Next, the optimum use of communications technology should be considered and any necessary changes identified. There may well be opportunities for improving, or even replacing, existing business processes or products and services.

The External Business Environment

Every business operates in its own unique environment with its own opportunities and constraints. This might include such issues as:

- *Communication facilities available to your suppliers and customers.* It would pay to know how many of your customers have email, web access, video telephony, EDI, electronic cash, or any other advanced services that you could make effective use of for marketing, sales, delivery (in the case of information and software), invoicing, and cash collection. This sort of information can be gleaned from surveys that you or third parties carry out. Customer attitudes toward new ways of communicating are of great importance. While many millions of consumers have access to the Internet today, the number of people who use this access to place orders for goods is very small. Time can change this, but only if there is a real benefit to the customer (e.g., lower price or increased convenience).

- *The willingness of suppliers or customers to adopt new systems.* It is often far from easy to persuade either customers or suppliers to make greater use of communication systems. In many cases they will have to be convinced of the advantage to them of using particular applications. Even where there are no such barriers, there can often be significant technical issues to be addressed. These include incompatibilities in communication systems and in other information systems. Often these difficulties can throw an entire initiative into question.

- *The use of the technology by existing, or potential, competitors.* Trade magazines can be of help here, as can other informal means.

- *Customer expectations.* Do our customers expect improved delivery times that can only be achieved through technology? What sort of information do our customers expect to get when they call us?

- *Existing and potential markets.* Communication facilities can give access to new markets (e.g., through the use of web commerce and/or call centers with international freephone).

- *Labor market.* Communications technology makes it easier for a business to move some of its operations to parts of the world that are less costly to operate from.

- *Existing and potential products.* Communication systems allow entirely new products to be created and existing ones to be enhanced. One exciting development is "mass customization" where volume-produced goods are tailored to individual preferences. Products currently available range from agricultural machinery to items of footware and clothing.

The Internal Business Environment

- *The information requirements of your employees.* This will give you an insight into what information systems or people they will need access to. It will also allow an assessment to be made as to just how critical continuous access to communication is (e.g., what happens if a link is lost for x hours?).

- *The mobility requirements of your employees.* Is it on site or off site? Will they require voice-only communication or voice and data when on the move?

- *Staff expectations and attitudes.* Will the staff adapt to new ways of working? Have they sufficient training? Have they proved adaptable in the past?

- *Core competencies.* Sometimes a business can suffer because although it has a competent staff, these staff members are in the wrong location and cannot easily be moved. Take, for example, a business that decides to centralize a part of its operation (e.g., customer support). Staff members in regional offices who already have most of the skills necessary for this work could be networked into a virtual call center and thus carry on working from their home town.

- *Organizational structure.* Is it assumed that the existing structure will continue unchanged or will proposals, calling for significant changes, be considered?

- *Organizational restrictions.* Organizational restrictions can occur in cases where the organization is geographically dispersed or has a number of business units operating autonomously. Alternatively, re-sponsibility for various communication disciplines (e.g., voice and data) may be delegated to different functional areas and may have to be catered for independently.

- *Budgetary restrictions.* What budgetary and financial constraints must be applied?

- *Scope.* The scope of a particular strategic initiative is often limited, either in organizational reach or in the range of requirements addressed.

The Communication Environment

- *Alternative service providers.* The number of service providers (carriers) is increasing dramatically in most countries. This is driving costs down while greatly increasing the range of options available. It is therefore important to take a medium-term view before locking in to any solu-tion that might limit the scope to take advantage of any new opportu-nities available.

- *Tariff trends.* See Chapter 16.
- *Regulatory environment.* How tightly regulated is the communications marketplace? Is further liberalization promised? If so, the likely impact should be assessed and due allowance made.

Communication-Enabled Goals

The following real-life examples illustrate how communications technology has benefited business in the recent past. They are presented here as answers to questions you can ask about your own business. How can communication be used to:

- *Reduce business costs?* By using videoconferencing to reduce expenses for travel and meetings. Communication systems can enable a business to shift the location of capital or relocate inventory. A well-known retailer reduced inventory costs by intertwining its network and business processes with those of its suppliers, helping it to implement a just-in-time approach to inventory management.
- *Improve operational efficiency?* By using ISDN to reduce the time taken to send large documents. By using groupware to allow review and interactive modification of a document by two or more people who may be in separate locations.
- *Meet customer requirements more effectively?* By providing requested information, as required, more efficiently than competitors.
- *Create product or service differentiation?* By using mobile communication to make product-support personnel more responsive. By using a dedicated data network to improve transaction speeds for a cash-dispenser service.
- *Provide added value for customers/suppliers?* A service organization increased market share by using its network to provide business partners and their customers access to critical information. The network can provide a competitive edge by linking the business processes of different organizations, permitting strategic synergy with customers or suppliers.
- *Reach new markets?* By advertising on the Internet, many companies have gained a significant improvement in international business.

13.3.3   Sources of Information

Useful sources of information about communication technologies and solutions include the following alternatives.

Trade Associations, Shows, and Magazines

Most industry or commercial sectors have one or more such organizations, many of which issue regular publications. These can give you information and ideas as to how your competitors are using technology to the best advantage. If you cannot be innovative, at least you can copy.

External Consultants

A good consultant can provide very valuable advice. The main difficulty is to identify good ones who have plenty of relevant experience to draw upon. You could ask for a list of recent clients and check out what they thought of a particular consultant. A good consultant's time is generally very expensive, so it pays to have read up on the subject beforehand and to have some pertinent questions ready. It will usually help to have two people meet the consultant so that maximum benefit can be gained.

Not all consultants are entirely independent and without bias. Some have informal or formal links with suppliers and will tend to suggest solutions that these suppliers can provide. Even those who have no such links will rarely be knowledgable about all possible options and are likely to have their own favorite solutions.

Service Providers and Equipment Vendors

Equipment suppliers and service providers can be a valuable source of information on potential products, services, or applications. They can help identify a range of options, some of which might hitherto be unknown. You can file a *request for information* (RFI), a formal means of acquiring such information prior to any tender. They can take the form of a written request or even an advertisement published in a magazine or journal. Vendors are less likely to give detailed or innovative responses to a RFI unless they feel they have a realistic prospect of winning a contract. This means that targeted, or apparently targeted, RFIs are much more likely to get a quality response. Most suppliers have been stung too often to give free consultancy with no realistic prospect of a return on their investment. As one supplier put it, they give the menu but not the recipe.

Communications Technology Magazines

There are many magazines dealing with communications technology. Some are aimed at service providers, while others have a very strong user focus. Such magazines often contain case studies and descriptions of new products and services. Many are published on the Internet, some with free access and others charging a subscription. The quality of magazine articles varies quite a bit. It is always prudent to expect mistakes and to double check the facts before making investment decisions based solely on such information.

The Internet

Nobody can deny that there is a vast amount of information available on the Internet—some of it useful! Internet searches can bring up all sorts of information about products, but they are generally quite time consuming. News groups are also an interesting source of people's real experiences with products and services. As with magazines, double checking is recommended.

Other Industries

Monitoring how other industries have deployed communications technology can be a useful source of information. They will also be much easier to approach than a competitor. A number of companies have successfully adapted ideas obtained in this manner.

This Book

Books on technology tend to be a little out of date by the time they reach their readers. We hope that this one will give you some useful ideas, but most of all we hope it will help you to understand the products and services that you find out about using these means.

13.4 Review

A communication strategy is not just a one-time communication plan—even though a strategic communication plan may be one of its deliverables. It is also much more than a list of developments to be put into place over some specific time period (typically one to five years). For maximum effectiveness it requires frequent reviews in line with changing business, technological, and regulatory environments.

Sometimes it will happen that during the planning stage, an idea will emerge that will cause you to have to rethink things that have already been decided. This is to be expected and should not be avoided simply because it will delay things. It should be remembered that mistakes or omissions are relatively simple to correct during the planning stage but can end up very costly if an unsuitable system is implemented.

Neither is it a linear process. It is a process of iteration encompassing all areas of the company. As new communication-enabled opportunities or goals are identified, these in turn affect other strategic processes, which in turn impact the communication strategy.

13.5 Scope

Ideally, any ICT initiative should be all encompassing and address the requirements of the entire business in an integrated fashion. Where this is impracticable, appropriate guidelines should be in place to assist the decision-making process. Such guidelines could be the responsibility of the communication manager or information technology manager, depending on their scope and the particular organizational structure. Final prioritization for large communication projects should also be done at the corporate level and include all projects requiring funding—not just proposals relating to ICT.

There are often significant drawbacks or benefits that are only visible from a business-wide perspective. For example, if different functional areas of a company buy a diverse range of hardware or software, this can result in significant additional costs. However, such costs may not even be visible to the individual areas involved. The net result might be that an option that appears to be more expensive or less favorable in a direct comparison with its alternatives might be the more viable choice when viewed at a global or corporate level. Notwithstanding this, the range of proposals and their anticipated impact are often too diverse for such an all-inclusive approach to be feasible.

The advantages of a business-wide approach to ICT, in large companies, include:

- *Corporate prioritization:* Communication projects should be prioritized, and funds allocated in terms of overall business objectives. This can only be achieved if a corporate approach is taken. Otherwise, the communication budget of the individual *business unit* (BU) is more likely to reflect the negotiating skills of the budget holder. Anomalies can then arise where some BUs have the resources to implement low-priority ICT requirements while others have to postpone more business-critical projects.

- *Economies of scale:* Suppliers should offer more favourable discounts than if each BU were to address their communication needs individually. It may also be possible to share communication systems (e.g., a LAN) between two or more areas. Potential savings here include initial procurement costs, support and maintenance charges, and a reduction in the number of software licenses necessary.

- *Standard solutions:* Often a small number of standard solutions will meet the communication needs of most BUs. Having such solutions allows support functions be streamlined and made more responsive. It also minimizes the risk of "communications islands" due to disparate applications, or even operating systems, being implemented by different BUs.

- *Service sharing:* Sometimes communication services (e.g., EDI, email, EFTPOS) are deployed by an individual BU, or group of BUs, to meet a specific need. It may well be possible to leverage extra value from these services if other BUs can find additional applications for them.

- *Communication systems efficiency:* Optimum communication solutions are possible only if all communication needs are considered jointly. This is particularly the case with intersite communication needs where the cost of connectivity is greatest and very high efficiency levels are desirable.

- *Improved process management:* Information systems are a key factor of process-based organization. Such businesses are critically dependent on the free flow of information between the various functional units. To achieve this all information systems must be able to communicate without impediment. This cannot be realistically achieved if each BU is free to implement individual solutions without having, as a minimum, to meet mandatory corporate communication guidelines.

Take a simple scenario that illustrates some of the foregoing advantages. *A large business is segregated into a number of business units. Each BU has a number of desktop PCs all of which are connected to the company WAN. These PCs are now being upgraded, as is the associated application software.*

In one company each BU has its own operating budget and is expected to independently choose the most suitable solution for its own particular needs. Some buy PCs from company X with a software suite from the market leader pre-installed. Others buy their computers from company Y citing advanced hardware features as the primary justification. These BUs buy their software suite from an alternative software developer because it has all the features they need and is significantly cheaper. The remaining BUs adopt a variety of approaches.

Each BU is confident that they have made the most cost-effective choice given their particular circumstances. However, when viewed at an overall business level, the picture is vastly different. Because of the relatively small quantities involved, the unit price paid for both hardware and software is far too high. Meanwhile, due to the diversity of the solutions, ongoing support is both costly and inefficient. Even worse, the different software suites are not fully compatible. This means that documents prepared with one suite do not import properly into the alternative products. As a result, most users still send documents in paper format leading to delays and other inefficiencies.

Another business takes an alternative approach. Here, each BU still has its own operating budget but must follow defined guidelines prepared by the ICT area. Only a limited range of hardware can be purchased without special approval and this is supplied to a standard configuration. A single software suite is used throughout the business with standard templates for all common applications. In addition, all company LANs have the same printing configurations and most users can use any computer available and still get personalized "online" facilities.

With this approach end users have fewer difficulties using the new applications, and any difficulties that do arise are quickly resolved. As a result, users are more comfortable with the technology and make greater use of it. This facilitates the identification and development of further ICT applications. The overall result is improved business efficiency and reduced ICT costs.

13.6 The Business Case

It is good practice to require all communication proposals to be supported by a business case (Figure 13.2). This ensures that only those proposals that further the overall business strategy are authorized. In the absence of such a policy, proposals may well be chosen merely because they are well written rather than on the basis of the value they add to the business. This is a particular danger with communications technology proposals, as they are typically presented from a technical perspective, which can be difficult for a nonspecialist to understand.

Figure 13.2 Typical business case.

The primary role of any business case is to facilitate the appraisal process, which is normally carried out before a proposal with any significant business impact is authorized. If, as would normally be expected, there is competition for resources, the business case is also used to prioritize the various proposals. To this end, it should include sufficient information to allow a full and systematic appraisal of the proposal. It should be couched in nontechnical language so as to be readily understandable by all interested parties.

To ensure that all business cases are thoroughly prepared, including a full consideration of possible downsides, it is strongly recommended that full accountability for results be assigned in advance. Where there is any significant deviation between the expected and actual outcome, the responsible party can then be asked to justify this. This approach discourages speculative proposals being presented in an overly optimistic or unrealistic manner.

The following section is intended to give an insight into some of the main issues covered by a typical business case, with particular emphasis on communications technology concerns. However, where an extensive financial appraisal is required, say, before final approval for a large project, a high level of detail, beyond the scope of this book, may be required. It is recommended that such detailed analysis be left to financial experts.

13.6.1 Time Span

The time span covered by a particular business case will have a critical impact on the assessment process. During the 1990s, such time scales have shrunk significantly, with some businesses requiring their plans to produce results in less than a year. As the individual business cases feed into the business plan, this means that they, too, must also have a prime focus based on similar time scales. In practice, this means that the time period for implementing a particular initiative must be minimized if results are to be achievable within an acceptable time scale. However, it is not always possible to keep the overall evaluation period for a communication initiative this brief, particularly where significant startup costs are involved. In such cases, an initiative may only be viable if costs can be written off over a number of years.

With communication networks, the optimum evaluation period often depends on the economic life span of the associated equipment. This is the duration, at the end of which it will no longer be financially advantageous to continue to operate the equipment. The economic life span is linked to the depreciation policy, which is the time period over which the cost of the equipment is spread for accounting purposes and the manner in which this cost is written off during this period. A typical depreciation policy would write off the value of communication equipment, in a linear fashion, over three to five years. In such cases, the evaluation period will typically be of a similar order. The general recommendation is to keep the overall time scales as short as possible, consistent with an acceptable rate of return on the investment required. In particular, it is recommended that time scales greater than five years be avoided, as any communication solution is likely to need reconsideration after such a period.

13.6.2 Typical Business Case

Executive Overview

The executive overview gives a brief resumé of the proposal and highlights points of particular significance. It is often used to determine whether a proposal warrants more careful consideration and may even form the basis for a decision.

Introduction

This section gives the background to the proposal and why it is required. It should also indicate the overall scope of the proposal and indicate what is and isn't being addressed. It should also indicate where the proposal fits in to the overall communication strategy.

Business Implications

Here, the proposal is given a strategic business context by estimating its impact in terms of one or more key business goals. The expected results should be tangible and, if at all possible, quantifiable. Associated time scales should be clearly indicated, and it is normal to draw a comparison with any existing solutions. Where a proposal is speculative, it should be clearly indicated and justified as such. It may also be appropriate to discuss how costs are to be apportioned, particularly where this involves a departure from current practice, such as service level agreements (see Chapter 14).

Many communication proposals will have direct business implications, which can be clearly stated. For example, an application such as CTI might be of such benefit to a telesales operation that it merits replacing an entire communication system. In this case, the proposal would highlight areas such as increased efficiency, more targeted selling, and reduced costs. Alternatively, there may be direct cost savings—for example, tariff savings from better utilization of communication links. In cases such as these, it is relatively easy to provide a strong financial justification for the proposal, which should mean that it is readily approved.

However, some communication proposals may have potential benefits that may not be readily quantifiable. Alternatively, it may be difficult to assign them a cash value at this time. It is important in such cases that the overall business value of the proposal be strongly argued. This ensures that the full potential of the system is recognized at the assessment stage and means that the proposal is less likely to get rejected solely on economic grounds (see the next section). For example, let's examine the proposal for a new information system for a retail chain. It will allow customer profiling, targeted marketing, and valuable information to the business and its suppliers. Ultimately, the strategic importance of this proposal may be immense. It is expected to allow a much-improved customer service, which will increase the customer base. Reduced stock holdings should be possible, as purchases can be tracked. Meanwhile the database information itself may have a significant intrinsic value that can be realized.

Financial Implications

Financial criteria are normally the key considerations when sanctioning, or prioritizing, a particular proposal, especially where significant capital expenditure is involved. As a result, the primary aim of an appraisal process is to ensure the financial viability of a proposal—often to the exclusion of all other factors.

Where possible, cash values, preferably based on market values, should be assigned to each expected benefit or cost. If any figures are speculative or a matter of judgment, this should be indicated together with an estimation of their accuracy. Care should also be taken that attributable savings can actually be cashed in. For example, a new switchboard might allow a single telephone operator to handle twice as many calls. It could be tempting to allow a 50% labor saving in such cases. However, such a saving can only be realized if a sufficient volume of calls can keep the telephonist fully occupied or additional work of some other nature can be assigned.

Alternative Options

There is often more than one means of satisfying a business' communication requirements, both from a technical and financial viewpoint. Indeed, there would be very few instances where there would not be a number of viable alternatives. From a technology perspective, for example, it might be possible to use either a frame-relay or ATM solution. Meanwhile, from a financial and management perspective, the advantages of outsourcing versus a wholly owned private network might need to be considered. For larger projects, where the overall cost justifies the additional workload, it is recommended that a number of different options be presented for consideration. This should include a minimalist option, where the existing infrastructure and associated services are utilized to the maximum extent possible. A recommendation as to which is the favored option, with full justification of the option being proposed, should then be given. For smaller projects, a number of potential alternatives should at least be identified, if only to show that the option being proposed has been fully thought out.

Risk Analysis

Risk analysis is used as a means of quantifying the effects of the assumptions underlying any particular proposal, varying over a given range. In the case of a communication systems proposal, there will be communication-specific variables, such as varying tariff costs and the rate of obsolescence of communication equipment. Another factor to be considered is the flexibility of the solution if, for example, additional communication services are required or a network topology has to be significantly altered. This latter situation could easily arise if additional business locations have to be included or existing locations are being moved or even closed.

Before the risk analysis can be carried out, the likely range over which each variable can vary must first be determined. The most uncertain variables are then chosen to give a worst-case scenario. Sufficient contingency should then be incorporated to allow for the worst-case scenario. At worst, particularly for speculative proposals, this might involve writing off the entire investment. In such cases, it would be wise to assess the risk of such an occurrence and its acceptability.

Appendixes

Additional information, particularly of a highly specialized nature, that might aid the decision process is best included in appendixes. Appendixes are particularly useful with complex proposals, where they can be used to limit the amount of information in the main body of the report. Typically they would include the following information:

- *Implementation plan:* Project time scales and significant milestones are shown here—often in bar or Gantt chart form.

- *Financial tables:* A detailed breakdown of expenditure in each category is given here.

- *Technical information:* It is normally unwise to include significant technical detail in the main body of a business case. Many individuals involved with the appraisal process do not have a technical background, and such detail may actually hinder their analysis. Conversely, those individuals with appropriate qualifications and experience may wish to evaluate the technical aspects of the proposal more thoroughly. By including the technical detail in an appendix, both requirements are satisfied.

References

[1] Porter, Michael E., *Competitive Strategy: Techniques for Analyzing Industries and Competitors,* New York: The Free Press, 1980.

[2] Hammer, Michael, and James Champy, *Reengineering the Corporation—A Manifesto For Business Revolution,* London: Nicholas Brealey Publishing Limited, 1993 in association with HarperCollins Publishers.

[3] Mintzberg, Henry, *The Rise and Fall of Strategic Planning,* New York: The Free Press, 1994.

[4] Manganelli, Raymond L., and Mark M. Klein, *The Reengineering Handbook,* New York: Amacom, 1994.

[5] Long, Carl, and Mary Vickers-Koch, "Using Core Capabilities to Create Competitive Advantage," *Organizational Dynamics,* Vol. 24, No. 1, Summer 1995.

[6] Coombs, R., and Richard Hull, "BPR as 'IT-Enabled Organizational Change': An Assessment," *New Technology, Work and Employment,* Vol. 10, No. 2, Sept 1995.

Further Reading

Kovacevic, Antonio, and Nicols Majluf, "Six Stages of IT Strategic Management," *Sloan Management Review,* Vol. 34, No. 4, Summer 1993.

Pine, B. Joseph, Don Peppers, and Martha Rogers, "Do you want to keep your customers forever," *Harvard Business Review,* March/April 1995.

Tapscott, Don, and Art Caston, *Paradigm Shift—The New Promise of Information Technology,* London: McGraw-Hill, 1993.

Valovic, Thomas S., *Corporate Networks—The Strategic Use of Telecommunications,* Norwood MA: Artech House, 1993.

14

Developing a Communication Solution

14.1 Introduction

This chapter concentrates on developing a communications technology solution rather than an overall information and communication technologies solution. It will be appreciated that, in practice, it is not possible to deal with communication systems in total isolation. In addition, many of the principles outlined apply to all ICT systems.

By now, the expected role to be played by information systems should be broadly clarified. Next, a communication solution must be developed to meet these needs. Depending on the particular circumstances, the optimum solution can vary significantly even though, on the surface, the requirements might be very similar. This means that developing an optimum communication solution is by no means easy. The first, and often most difficult, stage is to draw up a specification that details the actual requirements in a systematic manner. If this task is carried out successfully, business value will be the primary driver for the development of any proposed solution.

14.2 Communications Technology as a Service

Traditionally, communication costs have been treated as a business overhead. This means that their true cost, and business value, are hidden. This is particularly true for large businesses where individual *business units* (BUs) often ignore such overheads in their strategic planning. A better approach is to treat communication as just another service with costs and benefits. This ensures that business value, rather than expediency, drives the development of communications technology. How business-critical each service is can then be determined and be related to service expectations and cost of provision.

With larger businesses, the individual needs of the various BUs can be assessed individually. This has the advantage that the range of services required, and their associated *quality of service* (QoS), can be closely aligned to the actual needs of each BU. Such an approach allows business-critical BUs to be targeted for a particularly high QoS with minimal cost constraints and maximum in-house control. For other BUs, cost could be the key parameter in evaluating the various options.

The description of what constitutes a service is normally written as a formal *service definition*. Depending on the individual circumstances, these can vary greatly in their scope and level of detail. For internal use, a relatively simple description should suffice. As an example, the basic communication infrastructure (e.g., the LAN, PBX, PSTN, and ISDN) might be categorized as one service. Similarly, a number of standard applications (e.g., email and voice mail) could also be bundled as part of the same service. Additional high-value applications or facilities, such as videoconferencing or broadband Internet access, could be classified as separate services in their own right. The charging policy implemented will have a major bearing on this area.

14.2.1 Service Levels

A number of different levels of service are often defined to cater for a range of different business needs. Each BU or end user could then be given the basic service level by default and have to specify any enhanced levels required. From the service provider's perspective, this approach has the advantage that it is easy to identify who requires nonstandard, and often costly, service levels. From a business perspective, the service level provided is directly related to real needs, and it is easier to incorporate the actual cost of communication services into the final product or services.

Service Level Agreements

A *service level agreement* (SLA) is an agreement between a service provider and a client. It lists all pertinent details of the service being provided. It documents the specific concerns and obligations of each party and provides suitable redress if the agreed level of service is not provided. From a service provider's point of view, SLAs provide an equitable basis for charging users depending on the level of service they demand. With SLAs there is a greater incentive to maintain high standards, as the service provider is more fully aware of the business impact of a degradation in service. In many cases, suitable penalty clauses may also be invoked if a satisfactory service level is not maintained.

The primary basis for any SLA is the business-requirements specification as discussed in the next section. Other areas that may be included are:

- *User responsibilities.* It is not reasonable to expect the service provider to provide service guarantees under all circumstances. It is prudent to include a list of user responsibilities in any SLA. For example, support may be withdrawn if nonapproved applications are used. Software downloads from nonapproved Internet sites may be prohibited or subject to specified constraints. Installation of noncertified hardware may also be prohibited.

- *Service provider responsibilities.* All critical service provider responsibilities should be listed. This would include time frames for response and repair and the mechanism for resolving any service-level violations.

- *Third-party responsibilities.* Not all aspects of a communication system need be under the control of the service provider. In such a case, the responsibilities of any third party (e.g., an equipment supplier) should be stated.

- *Priorities of users/BUs.* In some cases (e.g., multiparty communications outages), it can be necessary to restore service on a prioritized basis. It can save a great deal of argument if the relative priorities are agreed in advance.

- *Escalation procedures.* A formal escalation procedure is often included to cater for ongoing difficulties or disputes.

All SLAs should incorporate a feedback mechanism to ensure that the service provider is meeting his or her obligations. In a communications technology environment, this feedback is by means of suitable performance parameters used to monitor the QoS being delivered. Choosing suitable performance parameters is not always a trivial task. The parameters chosen must be measurable, and there must be an agreed mechanism in place for making the measurements. Once chosen, individual performance targets must be assigned to each parameter. If these performance targets are not met, the service provider will be in breach of his commitments and may be liable for penalties.

When an SLA is being drawn up initially, it is wise to make provision for a review of both the parameters being monitored and their thresholds after a reasonable period. This allows both parties to fine tune the agreement in light of their experience. Such a provision should not be seen as an opportunity for either party to renege on their commitments. Rather, it should give rise to a final document that both parties are fully committed to. In the absence of such a provision, each party will adopt a conservative approach and endeavor to protect their own interests at all costs. An ongoing re-evaluation process should also be put in place.

14.2.2 Service Availability

The service-availability level is possibly the most important service-level parameter. This is defined as the proportion of the total time that the service is operating within specification. It can be expressed as a value between one and zero or as a percentage. For example, a call center might well require its telecommunication service to be available on a 24-hour basis, 365 days a year. If this service had a guaranteed availability of 0.9995 (99.95%), it shouldn't be out of order for more than four hours during any one year.

All services will have an intrinsic availability level depending on the particular circumstances, and there will often be a significant cost implication if a higher level is requested. It is therefore very important that an optimum availability level is specified. To ensure this, it is necessary to quantify the business impact of an information systems outage. The standard means of doing this is to consider the cost of downtime and to compare it with the incremental cost of increasing the availability level.

Ideally, the relevant business manager should carry out any calculations, though, in practice, the task often falls to the communication manager. A useful starting point is to consider the hourly costs of an outage (both direct and indirect) for each area of the business. This will include employees' time wasted and more importantly lost business and loss of customers. A general global calculation should then be made to set an optimum baseline availability figure for the business. However, particular care should be taken to separate out business-critical areas, where the high costs of downtime will justify higher spending on increased availability.

Measuring Availability

The availability of a fixed connection (e.g., a private line) is generally measured in terms of *available time* (AT). Such a link is considered to be available if the maximum number of errors per unit of time does not exceed a given level over a specific time period. The idea is that during AT, all applications should work within specification. Many service providers can now produce AT statistics on request.

Measuring the availability of a switched connection, or comparing it to that of a fixed connection, can be problematic. The ITU-T (I.355, 1992) has introduced the concept of usability for determining the availability of ISDN services. *Service availability* (SA) is defined as the percentage of time that a usable call can be established and maintained between two end points. There is still a difficulty in agreeing what exactly a *usable* call is!

Availability can also be measured in terms of reliability and serviceability. The reliability or *mean time between failures* (MTBF) is a measure of how often a breakdown occurs. The serviceability or *mean time to repair* (MTTR) is a measure of how quickly normal operation can be restored after a failure.

$$\text{Availability} = \text{MTBF} / (\text{MTBF} + \text{MTTR})$$

14.2.3 Charging for Services

A major advantage of a service-oriented approach is the ability to charge users and, where appropriate, external customers for communication services. Charging for services means that a direct link is established between communication facilities and cost. It helps ensure that all requests for services are considered in terms of the added value they bring to the business area involved. It is far easier to justify an additional budget for new services if they can be shown to be cost effective and give a clear business advantage. Likewise, unwarranted demands for new services or unrealistic service level expectations are more easily curbed if the user or BU has to justify the associated expenditure or even meet it from their own resources. The real challenge is to strike a balance between discouraging excessive user demands while encouraging appropriate use of ICT to the fullest extent possible.

Charging Mechanism

A range of mechanisms, ranging from the simple to the very complex, can be used for charging. One approach is to limit the number of individual services while offering only a limited number of service levels for each. A charging policy based on the actual services provided with a premium for enhanced service levels could then be implemented. Costs could be allocated on the basis of the number of users or, where practicable, on actual measured usage. Whether or not all communication costs should be distributed in this fashion must also be considered. It could be argued that some costs (e.g., ongoing investigation into the role of ICT) should more properly be considered to be overall business overheads.

Internal billing can be on either a nominal or formal cost-transfer basis. At the time of writing, few off-the-shelf billing systems were available that could deal comprehensively with the many elements of a typical communication solution. As a result there is a significant tradeoff between the amount of billing information available and cost. This means that a pragmatic approach, sacrificing accuracy for simplicity, is often best, particularly for internal applications.

Implementing a billing system can be more of a challenge where the ultimate objective is to pass on communication charges to customers.

14.3 Specifying Communication Requirements

The first step in specifying communication requirements is to reach an agreement on the required services and their associated service levels. The next step is to draw up a requirements specification listing all issues of importance relating to each service. This document will be the framework around which the entire procurement process will subsequently revolve. It should couched in nontechnical language and be readily understandable by all interested parties. Any unnecessary detail should be avoided consistent with minimizing the risk of ambiguity. The end result should be a document that clearly indicates what each user or BU is trying to achieve.

Because of its pivotal role, it is important that all stakeholders are involved in formulating the initial requirements specification. They should also understand and be in broad agreement with its main thrust. Who these stakeholders are will vary according to the cost, potential impact, and strategic role of the procurement. It is recommended that end users be always classified as stakeholders and their representatives included in the specification process. Not only can they make a valuable contribution to the overall process, their inclusion also helps ensure that they buy in to the project at an early stage.

14.3.1 Writing a Business Specification

A business specification considers the communication requirements from a nontechnical perspective. Its central component is a list of the required services along with any associated features or facilities. Ideally, these requirements should be as generic as possible to facilitate a wide range of solutions. However, where particular features are only supported by a specific technology or require a vendor-specific product, this should be clearly stated. All requirements should be classified in terms of core functions, support functions, and nonessential functions. The business importance of each should be documented as well as the service-level expectations during normal and failure conditions.

An attempt should also be made to consider ongoing requirements. This would include the likelihood of new applications, particularly any that may be bandwidth hungry (e.g., multimedia). The expected rate of business churn is another issue. Areas to be considered include expansion plans, proposed site closures (or openings), or an intent to move to a new location. Any known constraints must also be identified. For example, there may be legacy systems that any final solution must support, or a particular public network service might not be available at all locations in a multisite network. Physical restrictions may arise due to the layout, age, or location of a building. Time constraints (e.g., for project completion or for change requests) may also arise.

Once the business specification has been written, the next step depends on the individual circumstances. One option is to use this document, without further elaboration, as the primary basis for the procurement. In some instances, this approach is perfectly valid, especially where the costs involved are relatively small. However, for larger procurements, it is recommended that a formal technical specification be written (see Section 14.4).

14.3.2 Optimal Business Solution

Technical merit, on its own, is not sufficient to guarantee an optimum business solution. It is recommended that any final solution be assessed using the following criteria:

- *Support facilities.* Full support facilities should be available at all locations, including remote sites and for mobile users. The extent of the support, and the conditions under which it will be given, can determine the viability of an entire communication solution.
- *Report capabilities.* A range of statistics should be available to allow all aspects of the communication system be properly managed. Ideally, periodic reports covering billing, network utilization, and actual service levels achieved should all be available.
- *Level of engineering.* Maximum cost efficiency will only be achieved if a solution is neither over- nor underengineered. At the implementation stage, overengineered designs that add excessive costs should be avoided. Later on, network utilization reports should help highlight any serious deficiencies.
- *Business change management.* The business environment is always changing and the degree to which this change can be accommodated should be critically evaluated. There might be expansion or contraction within the business. New sites might need to be added or existing ones moved or even closed. New partnerships, takeovers, alliances, or even something as mundane as new suppliers may all affect the communication requirements.
- *Degree of lock-in to suppliers.* Is it possible to change suppliers without excessive cost? Alternative suppliers may well be offering attractive alternative packages at a later date. If there is a substantial degree of

lock-in, this should be justified on technological or commercial grounds.

- *Degree of lock-in to technology.* The relentless march of technology continues, bringing new capabilities, new communication services, and new opportunities. It should be possible to take advantage of these opportunities at minimum cost and disruption.

- *Evolutionary strategy.* A clear evolutionary strategy should be an integral part of any communication solution. This accommodates forecast advances in communications technology. Any technology strategy must facilitate unforeseen technological developments that may need to be considered at minimum notice.

14.3.3 Advantages of Standardization

A high degree of standardization in both applications and technology is often required if optimal use is to be made of information systems. Due the relatively low cost of many systems, spending authorization is now often at relatively low levels in the business—sometimes even at individual user level. In this environment, formal guidelines can ensure that the majority of deployments are compatible with the overall information systems strategy.

Some of the advantages of a successful standardization policy include:

- *Maximum connectivity.* It should be possible to freely exchange information both within the company and, where appropriate, with suppliers and customers. In particular, there should be no unintentional information islands with which it is difficult to communicate. Success in this area can have exceptional business impact by facilitating advanced applications such as data warehousing or companywide intranets. At a more mundane level, initiatives such as companywide print on demand have no less significant impact.

- *Lower costs.* Significant discounts can often be negotiated where information systems are standardized, particularly where it facilitates bulk purchasing. Support costs can also be reduced.

- *Improved service levels.* With a limited product set to maintain, support staff can offer a more expert and speedy service. Maintainability is also enhanced because it is more cost effective to maintain spares facilities to address system failures.

- *Enhanced flexibility.* Suitable guidelines should ensure that there is maximum flexibility within the overall communication solution.

- *Greater security.* Standardization allows security concerns to be more readily addressed, as uniform facilities, such as audit trails or encryption, can be implemented in all appropriate areas of the business.

It is not always easy to enforce standardization guidelines. One approach is to refuse to support nonapproved systems—or even to refuse to connect them to any global systems. An alternative approach is to allow some local diversity but with the associated risk also being carried locally. This approach, while discouraging nonstandard solutions, still allows for innovation and avoids the danger of lock-in to a nonoptimal solution. Should a particular local initiative prove successful, it can then, if appropriate, be adopted more widely.

14.4 Specifying Technical Requirements

Converting business requirements into a technical specification is a skilled task, and it is recommended that experienced help be enlisted. As part of this analysis, it is normal to specify key QoS criteria in a standard fashion. The actual QoS parameters chosen and their values depend on the applications for which the network is being used and how business-critical those applications are. The more common QoS criteria associated with communication systems include the availability figure discussed earlier and the performance criteria described in the next section.

14.4.1 Network Performance Criteria

The performance of a communication network is a measure of how effectively the various applications, whose information it carries, operate. A range of performance criteria is available depending on the particular application. Some of the more popular ones are as follows.

Time Allowed/Time Sensitivity

This is a measure of how quickly a given amount of information must be transferred. It may be given as a specific amount of time or, alternatively, as a predefined period during which the information transfer must be completed. It is normally specified for applications, such as backing up files to a remote site, where a specific amount of information is involved. It is useful in situations where the information can be transferred during periods when the communication link is otherwise underutilized. This is a particularly useful parameter as it allows low-priority traffic—which can be sent at off-peak times—to be identified.

Information Transfer Rate

This is a measure of how quickly information can be transferred on an ongoing basis. This is often specified in terms of bandwidth in bps. Other measures may sometimes be more appropriate, not least because they are more user friendly. Typical examples include the number of database records that can be transferred per minute or the number of pages of newsprint that can be transferred in an hour. In all cases, the transfer rate will ultimately have to be converted to bps.

Response Time

The response time is an indication of the delay between a specific input and a response. It is often specified for interactive applications that are similar to *electronic conversations.* A typical example is a hotel or airline reservation system, which involves a number of question-answer interactions. Here a round-trip delay of two or three seconds, per query, might be acceptable. For many applications the response time is more critical than a network's ability to carry large amounts of information.

Number of Simultaneous Calls

This parameter can be used with any switched network. For example, it can be used with a PBX network to allow the capacity of intersite links to be calculated.

Voice Quality/Video Quality

Parameters such as these are more subjective and no definitive measurement standards exist. Some have been proposed but none have gained universal acceptance. Subjective parameters are often necessary at the requirement-specification stage, but they are seldom monitored on an ongoing basis.

Other Parameters

Many other performance criteria (e.g., the number of simultaneous sessions that can be handled by a particular LAN application) are in use.

14.4.2 Impact of Technology

A key business decision is when to migrate to a new product or even to a whole new technology. New facilities and features are constantly being offered and businesses can be afraid of being left behind. On the other hand, further enhancements are always promised for the future, and these might add even greater value.

Technology Life Cycles

There is a distinct life cycle associated with technological products and services. This can be broadly divided into three main stages:

- Leading-edge technology, where major development is still ongoing, often with minimal standards in place;

- Mature technology, where all major developments are completed and standards clearly defined. Mature does not, necessarily, imply old. The rate of development is now such that some communication technologies are now mature in less than one year!

- End-of-life technology, where all development work has ceased and vendor support is minimal or even withdrawn.

The next section gives some guidance on this issue, though, obviously, the optimum choice in any case will depend on the particular circumstances.

Leading-Edge Technology

First is not always best where new technology is concerned. The latest technology is often overpriced and rushed to the market to gain a lead on the competition. It is recommended that leading-edge technology should only be considered where it offers significant, business-critical advantages not otherwise obtainable. Ideally in such cases, a small-scale deployment should be made prior to any significant commitment. In addition, a fallback position should be considered in case the technology fails to deliver the promised benefits or even fails to work properly! For example, a large sales organization might be considering an Internet VPN as an alternative to toll-free (freephone) numbers for their traveling sales force. However, they have performance and security concerns regarding the viability of this alternative. A limited trial can be used to investigate these concerns with negligible business risk.

Should a decision to widely deploy leading-edge technology be made, a number of concerns must be addressed. One of these is the availability of suitable expertise, which is often at a premium where new technology is concerned. Another is the level of investment involved and the risk of having to write it off should the technology fail. In this regard, a particular consideration is the upgradability of the proposed solution. Most vendors will make promises regarding upgrade paths, but, in many cases, these never materialize or else do not apply to your particular model. In other cases, the cost of such upgrades can render them non–cost effective.

New technologies are much more likely to require upgrades than mature technology for the following reasons:

- Many products will have "bugs" due to the ongoing development work.

- Standards are often incomplete, leading to some incompatibility between equipment from different vendors (e.g., many of the early implementations of the various modem standards did not fully interwork).

- Enhancements are often required to allow all the functionality of the final standard (e.g., some early models of GSM mobile phone cannot transmit data).

- Leading-edge technology is more vulnerable to early obsolescence from a competing service or product.

Mature Technology

In most cases, mature technology offers a better business solution than leading-edge technology. Standards should now be agreed and any incompatibility issues, where applicable, solved. Products should be fully supported and shipping at greater volume at significantly lower cost. Vendors will now be concentrating on adding additional value-added functionality in an attempt to differentiate themselves from the competition. The net result should be a more attractive, reliable, and cost-effective product. For example, V.34 modems dropped in price by over 50% in less than one year while adding increased functionality, such as voice mail and CTI capability.

End-of-Life Technology

Many businesses will have older technology that still meets their requirements. In some cases, it can make economic sense to continue to operate this for as long as possible. This is particularly true where existing technology meets the main business requirements at a fraction of its replacement cost. However, before committing to this approach, a number of issues need to be considered:

- Support costs are likely to be higher than for more modern equipment. Much development work is now concentrated on adding functionality to reduce the level of support required.

- Availability is likely to decrease as downtime for repairs and maintenance increase.

- Maintenance costs are likely to rise as the probability of equipment failures increase. Meanwhile, spares availability can become problematic as vendors withdraw their support.

- Migration to newer technologies can become increasingly difficult and more expensive.

- Many advanced features, which might add significant business value, may not be available.

A Technology Strategy

The impact of technology should be a key consideration in developing the communication strategy discussed in Section 13.1.3. In particular, the role of technology in light of ongoing business requirements should be considered. This may well facilitate a clear evolutionary strategy for communications technology where the future role of specific technologies and services can be mapped out. For example, taking ISDN to the desk, as distinct from using it for PBX connections or LAN interconnection, might be the next evolutionary step.

As a minimum, it should be possible to identify the optimum use of technology at the current point in time. This might mean replacing existing communication equipment or changing to a different public network, even though the existing solution provides all the required communication services. Alternatively, a proposed change may be inadvisable at the current stage of development of the new technology, and it may be prudent to postpone deployment for a limited period.

14.4.3 Future Proofing

A communication solution must incorporate some element of future proofing if it is to have a reasonable economic life span. This means that it must be sufficiently flexible to cater for a wide range of additional requirements, which are currently unforeseeable. While no solution can be completely future proof, much can be done to minimize the risk of premature obsolescence. Some concerns, such as the license conditions for software applications, are general in nature and are addressed in Chapter 15. Others are directly affected by the particular technical approach chosen.

When considering how future proof any particular solution is, several areas merit particular attention.

Scaleability

How easy is it to provide additional communication capacity in response to rising demand? This growth may not always be forecast in advance. There might, for instance, be an unexpected increase in the number of PBX extensions required. One approach to providing scaleability is to overdimension the initial solution. This might mean installing a bigger PBX to allow sufficient spare capacity. A better solution, if it is viable, is to install a system that can be extended in a modular fashion.

Upgradability

Can equipment be readily upgraded to add new features or take advantage of new technology? Some equipment vendors are committed to the continuous evolution of their current products and offer specific upgrade paths. Limited upgradability is often possible through software upgrades. Other equipment can be substantially upgraded by changing specific hardware modules. Relying on upgrades can risky if you know neither the cost nor the effectiveness of the upgrade in advance. The previous track record of the supplier or manufacturer can give a useful indication of the reliability of any promises.

Flexibility

The overall flexibility of the solution is a good indicator of how future proof any solution is. The greatest level of flexibility is provided with modular equipment where entire configurations can be changed if necessary. At a less extreme level, the capability of equipment to be configured for as many parameters as possible should be investigated. This can be a lifesaver at a future date even though many of the parameters may appear useless or meaningless at the present time. The ability of the network nodes to handle all major protocols with the same cards, or with as few cards as possible, is another consideration. The use of standard interfaces, to facilitate interconnection with other vendors' equipment, should also be noted. Another aspect of flexibility is the capability to reconfigure existing nodes or add new network nodes.

Cabling

The cabling practice used will have a major bearing on how future proof any solution is. It can be extremely disruptive and expensive to have to install additional wiring at a later stage. This means that flood wiring a building, using a structured wiring system, is often the best solution to cabling requirements. To minimize the need for ongoing cabling work, the authors recommend that the highest quality cable available (e.g., category 5 shielded cable, at the time of writing) should be used for all such installations. The increased initial cost is more than compensated for by the additional flexibility afforded by this approach.

14.4.4 Availability Considerations

High availability figures are often specified in SLAs. It is often useful to have a general understanding of how this figure might actually be achieved. This can help determine the most cost-effective means of increasing availability. It can also be useful when assessing if availability figures being quoted by suppliers are aspirational or can actually be met. With external suppliers this can be especially important, as any penalty invoked is unlikely to fully compensate for a major communication outage.

Elements in Series

A typical communication system consists of a number of different elements connected together in series. Here, the overall availability is obtained by multiplying together the individual availability figures of each element. This setup is a good example of the old saying that "a chain is only as strong as its weakest link." In other words, what is required is a balanced system. It is no use paying a fortune for very reliable communication equipment if the availability of the communication link is very poor. Conversely, negotiating a very stringent SLA for a communication link is a waste of time and money if the communication equipment keeps failing.

Increasing Availability

Even the best and most expensive equipment will fail. Similarly, communication links will go down, regardless of the best efforts of the service provider. One means of increasing availability is to ensure that any faults are quickly rectified (i.e., decrease the MTTR). The downside of this approach is that availability figures can be misleading if the communication system is prone to frequent, short duration, outages. Consider a PRA ISDN line that is prone to glitches, each lasting less than one second. Even if such glitches occur on a daily basis, their effect on the overall availability figures will be negligible. However, each occurrence might cause up to 30 calls on the link to fail, causing significant business disruption. This means that it is wise to set a limit on the maximum number of acceptable outages over a particular period of time. Agreement, on what constitutes an outage, and how it is to be identified, should also be reached.

An alternative means of increasing availability is to increase the resilience of the system by adding redundancy. This might mean adding additional backup cards to an equipment shelf or installing an *uninterruptible power supply* (UPS) to protect against a power outage. Ideally, no single fault should cause an information systems outage. In practice, this goal may prove too expensive to implement. However, a useful increase in availability can often be implemented at minimum cost. This could be something as simple as using an ISDN connection as a cost-effective backup for a leased line or connecting adjacent PCs to different LAN segments to increase resilience.

Availability of WAN Links

Service providers use a number of options for increasing the availability of individual WAN links. Similar techniques are also used to increase the effectiveness of backup communication links. Full physical diversity, with two physically distinct routes through the network, can provide very high levels of availability. However, providing this level of redundancy may involve significant expenditure on the part of the service provider. In some cases, this cost cannot be shared with other users and must be met in full by the business. Partial diversity offers a less expensive alternative. Here, for example, two local access cables may be available, but they might share a common duct over some of their length. The disadvantage is that both cables could be damaged simultaneously and cause an outage.

Where redundancy is built into a network, it is often necessary to reconfigure a network to restore service under fault conditions. This can be done manually (e.g., in response to an alarm condition) or automatically. The latter is by far the most desirable option, as it not only causes minimal disruption to end users but also avoids the human factor and its associated risks. *Network management systems* (NMSs) are often used by service providers to provide automatic restoration of WAN links. These may have the further advantage of facilitating switching of traffic to a *hot-standby* site should information systems fail at the primary site. For critical applications, it is worth inquiring whether the service provider has a disaster-recovery plan (e.g., a duplicated management system). Without this, a single event could wipe out both your site and the NMS that should be able to switch over your circuits. It can also be useful to ask how diversity is tracked to ensure that it isn't compromised during network reconfiguration.

Alternatives to diversity are also utilized to reduce the possibility of a communication outage. Avoidance is where it is specified that the physical route avoids a particular geographic area. This might be near a location (e.g., a construction site) where the likelihood of a cable break would be particularly high. Radio-based communication links can often be a more reliable alternative than physical cables. Some earlier microwave systems had a poor reputation, but modern digital systems are very reliable—and cannot be dug up.

14.4.5 Flexible Access Systems

Many communication solutions require a number of different services at a single site. Often these are provided in an uncoordinated manner—though one service provider may be responsible for all services. Even where service provision is coordinated, each service will generally be treated as a separate entity with its own associated equipment. A typical approach is shown in Figure 14.1. It can be readily seen that even relatively low-volume users can end up with quite complicated communication solutions.

An alternative approach, as shown in Figure 14.2, is where the service provider uses a *flexible access system* (FAS). Here, all communication services are presented to the customer using a standardized managed platform and,

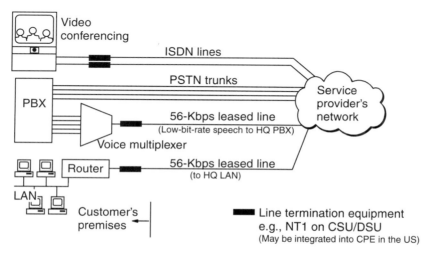

Figure 14.1 Traditional access to multiple services.

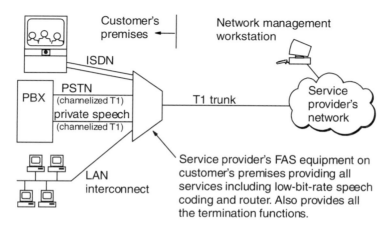

Figure 14.2 Flexible access to multiple services.

typically, a common access link. Some FASs, as in the case illustrated, also incorporate functionality more often associated with businesses' own CPE equipment.

FASs allow service providers to offer integrated solutions that increase their responsiveness to changing business needs. Communication services can be provided, reconfigured, or ceased—all from a remote management terminal. This allows very fast response times with requests for change often being accommodated in less than 24 hours. FASs also facilitate standard availability levels for all services. In particular, the benefits of any measures introduced to enhance the availability of one service (e.g., local loop diversity) can be readily shared with all others.

Comprehensive service monitoring is an integral part of most FAS solutions. This means that a wide range of statistics is available, which are often incorporated into SLAs. Some FAS solutions also allow businesses to have their own management terminal. As a minimum, this allows alarms to be monitored and statistics viewed in real time. More advanced systems also allow businesses a significant degree of individual control that may well extend to full service reconfiguration on a pre-arranged basis.

FASs can form the basis of extremely cost-effective communication solutions. There may well be tariff reductions as suppliers pass on savings arising from integrating the various services. (In practice, any savings are often offset by the cost of the service management.) However, even if the service provider charges actually increase, the cost of the overall solution may still be attractive. It is often possible to reduce the amount of CPE at each site, which gives significant savings. A business can also enjoy the functionality of a network management system while enjoying the economies of scale offered by a shared solution.

While FASs merit serious consideration, they are not always the optimum solution. In particular, they can mean that a business is effectively "locked-in" to a single service provider. While, in theory, it is possible to change suppliers, doing so can be problematic. This is particularly true if functionality is being provided by the FAS that would otherwise require external CPE.

14.4.6 Technical Specification

The business specification discussed in Section 14.3.1 considered communication requirements from a nontechnical perspective. As such, the level of detail included may not be sufficient to ensure that all technical requirements are identified. This need for detail is greatest where there is a high level of complexity, where specific legacy applications or equipment must be accommodated, or where a high level of control over the final solution is required. The requirements may also be specified in a manner that makes it be difficult to measure a supplier's compliance at a later stage.

A technical requirements specification considers the communication requirements in greater depth and from a technical perspective. The overall objective is to specify the requirements in such a manner that they are clear, unambiguous, and demonstrably verifiable. The necessary level of detail to achieve this can vary greatly. Where a business has a detailed technology strategy, the specification may require particular technologies (e.g., ATM) to be used or maybe even equipment from a small number of manufacturers. More usually, the specification concentrates on connectivity requirements and associated performance criteria.

A technical requirements specification can be considered to have two distinct elements dealing with functional and nonfunctional requirements, respectively. The functional requirements would typically include any onsite or intersite services to be provided along with any associated constraints. There might be specific applications to be supported or even specific protocols or equipment types. The level of detail may even extend to the specific interface types to be supported. Connectivity requirements may specified in terms of bandwidth and performance criteria required. Alternatively, only the applications to be supported may be given.

Next, a range of nonfunctional requirements must be addressed. Typical areas to be considered include:

- *Security.* Any security considerations must be identified at this stage, as it can be difficult and expensive to enhance security at a later stage.

- *Certification criteria.* An essential component of any requirements specification is the certification process that will be used to ensure compliance. This process is required to verify a solution after implementation and also for *acceptance testing* third-party solutions. Interoperability tests between specific applications, networks, or equipment often form part of certification criteria. This might be something as simple as an interoperability test with a supplier for an EDI connection. Conformance to specified standards is another option. This can have the advantage of being verifiable by independent tests.

- *Legal requirements.* Any known legal requirements should be listed. These might include safety requirements or type approval requirements. It can be prudent to include a clause making it the responsibility of the supplier to ensure that all legal requirements are met.

14.5 Communication Outsourcing

Outsourcing is a means of delegating responsibility for some aspect of an information system to an outside agency. It need not be an all-or-nothing scenario nor, necessarily, a large-scale operation. Indeed, it could be argued that most companies already outsource their communication requirements to some degree, as few own or control their own WAN links. The increasing intelligence and functionality of public networks, allied with advancing deregulation, mean that the number of outsourcing options seem likely to increase rather than diminish [1–9].

Examples of communication outsourcing include VPNs and Centrex, both of which have been discussed previously. Another possibility is managed services, where the end-user applications are also on a shared platform. Typical examples include public web servers, public LAN servers, and public notes servers. An alternative scenario is where services are provided over a value-added network. Here the value-added network is a shared public platform, but the service provider is not, necessarily, its owner.

14.5.1 Why Outsource?

Operating a private network requires significant financial resources to be tied up in the operation and maintenance of rapidly depreciating communication facilities. On the other hand, outsourcing is a pay-as-you-go means of meeting some or all of a company's communication needs. As such, it can free up significant capital resources and result in a greatly improved cash flow. There can also be significant savings due to the reduced overheads associated with an outsourcing solution. These include greatly reduced staffing levels, with the consequent reduction in associated costs (e.g., payroll, skills updating), and savings in ancillary facilities. This is not to imply that service providers are a zero-cost option. However, their economy of scale means that they benefit from increased leverage when negotiating with their suppliers and also allows them to operate with improved staffing ratios. The net result, in theory at least, is a lower cost operation with some of the savings being passed on to their own customers.

Cost is not the only reason for considering outsourcing. It could be argued that operation of private communication facilities is an unnecessary distraction for most businesses. As performance is contractually guaranteed, or at least should be, the role of inhouse management can be reduced to that of a strategic and watching brief. Also, many businesses lack the resources to deploy communications technology to maximum advantage. As a result, the potential for using communication to increase competitive advantage often goes unexploited. As communication is the service provider's core business, they should have the right mix of products and personnel to maximize the business benefit from a company's communication solution. The service provider is likely to have far more sophisticated equipment, in a shorter time frame, than that which could be economically justified for a private network. Their experience may also be invaluable in dreaming up innovative applications that might otherwise go unrecognized.

Outsourcing can also provide a degree of future proofing and hence provide some protection against loss due to premature obsolescence. This is always a risk with privately owned equipment, where the upgradability of even the most advanced solution is often limited. Many service providers, meanwhile, upgrade their equipment regularly either as part of their marketing strategy or as a result of agreements with their own suppliers. However, outsourcing may not provide the degree of future proofing promised. For example, in the United States, local state regulators limit the amount of capital investment that the local carrier may recover through its public rate base. This may impede the introduction of new technology or hinder the timely upgrading of existing equipment.

One of the difficulties with private networks, especially those based on leased lines, is that they can be very rigid and inflexible. In fact, lead times for leased lines can be as long as three months in some European countries. With appropriate outsourcing solutions, a network can be quickly reconfigured to suit changing demands. In addition, sites can be readily added or removed as needs dictate.

14.5.2 Arguments Against Outsourcing

Outsourcing is not a quick-fix solution to existing problems, nor can it be used as a substitute for an overall communication strategy. Indeed, if the existing systems are not optimized to meet business needs, outsourcing might even make matters worse. The level of risk associated with outsourcing can also be underestimated. For example, the existing communication services may add unique value that cannot be readily duplicated in an outsourcing environment. Alternatively, the service provider may not deliver the promised service levels or there may be unexpected levels of disruption during the changeover period. Any standard penalty clauses are unlikely to cover the business impact of such eventualities.

Where existing inhouse operations are being outsourced, there can be significant associated costs. Restructuring costs can occur if existing employees must be redeployed or made redundant. Significant capital may be tied up in existing equipment, and the cost of writing this off may be excessive. Software licenses for essential applications may not be transferable to a third party at reasonable cost.

Should an outsourced operation prove troublesome, it can be difficult to revert quickly to an inhouse operation. Equipment may need to be reinstalled, staff retrained (or even rehired), and applications reloaded. This process is likely to be time consuming, costly, and disruptive to business operations.

14.5.3 Reaching a Decision

The viability of an outsourcing solution will vary widely from business to business. It will depend on issues such as the resources, skills, and efficiencies available inhouse and how fully integrated any existing communication systems are with the critical business processes. In some instances, the apparent advantages of outsourcing will be obvious, while in others, it may be totally nonviable. Where the results are encouraging, the possible benefits of outsourcing need to be identified together with a list of requirements necessary for them to be achieved. The possible outsourcing options must then be identified and evaluated. Any evaluation should make due allowance for any benefits that would accrue from increasing the effectiveness of any existing inhouse solution. Failure to do this could result in the case for an outsourcing solution being overstated.

Once a decision to outsource has been reached, and the decision implemented, some inhouse technical expertise will still be required. First, the contract will have to be managed to ensure that all contractual obligations are being met and users needs are being fulfilled. This will require a least one manager with a technical background, who is familiar with both the requirements specification and the SLA. Second, strategic communication requirements will have to be constantly reevaluated in light of changing business needs. Third, the overall outsourcing package will need to be compared to competitive offerings on an ongoing basis. Finally, expertise will be required to renegotiate existing contracts, evaluate new suppliers, and negotiate new contracts.

References

[1]   Cross, John, "IT Outsourcing: British Petroleum's Competitive Approach," *Harvard Business Review,* May-June 1995.

[2]   Lacity, Mary C., Leslie P. Willcocks, and David F. Feeney, "IT Outsourcing: Maximize Flexibility and Control," *Harvard Business Review,* May-June 1995.

[3]   Strassmann, Paul, "Outsourcing: A game for losers," *Computerworld,* August 21, 1995.

[4]   Spee, James C. "Addition by Subtraction—Outsourcing Strengthens Business Focus," *Human Resources Magazine,* March 1995.

[5]   Heywood, Peter, "Global Outsourcing: What Works, What Doesn't," *Data Communications International,* Nov. 1994.

[6]   Wallace, David, "Who's Afraid of Outsourcing?" *News & Views,* Feb. 1994.

[7]   Lippis, Nick, "2001: The New Public Network," *Data Communications International,* Nov. 1994.

[8]   Gareiss, Robin, "Carriers Set Their Sights on Managed Services," *Data Communications International,* Nov. 1994.

[9]   Krone, Roger A., "Letters to the Editor," *Harvard Business Review,* July-August 1995.

Further Reading

Hart, Christopher W. L., "The Power of Internal Guarantees," *Harvard Business Review,* Jan.-Feb. 1995.

Venkatesan, Ravi, "Strategic Sourcing—To Make Or Not To Make," *Harvard Business Review,* Nov.-Dec. 1992.

15

Supplier Selection and Management

Changing regulatory environments combined with rapidly evolving technology mean that there is an ever-increasing range of communication products and services from which to choose. A side effect of this is that existing equipment and services are becoming obsolete sooner and hence need to be reviewed or replaced more often. Meanwhile, businesses are finding many new applications for ICT. This is fueling the demand for new communication services and means that businesses are becoming ever more reliant on communications technology for the success of their business. The net result is that the selection of suitable suppliers and their ongoing management is now a critical issue with significant associated risks.

This chapter builds on the previous chapter where the issues of identifying companies' requirements and writing requirements specifications have been addressed. Though not explicitly stated, the material leans towards communication services rather than equipment. It is assumed that potential suppliers will be furnished with a business requirements specification (see Chapter 14), for a large contract and be free to suggest their own technical implementation. Such an approach has the advantage of allowing innovative proposals from potential suppliers unfettered by any unnecessary restrictions. The material will need adaptation to meet individual circumstances, even though the general principles introduced should still apply.

15.1 How Many Suppliers?

In an ideal world, a single supplier would supply all of the communication needs of a company—providing the best service at the lowest cost. The nearest one can get to that ideal today is to outsource all of the communication requirements to a single service integrator, who will manage all aspects of the communications technology. However, this is not a realistic option for most businesses, and multiple suppliers are still the most common scenario. Typically, there will be one supplier for offsite links, one for data equipment, and one for voice equipment.

The procurement process should be used to assess the strengths and weakness of a range of potential suppliers. Each supplier will have access to different technologies or offer different services. Their range of technical expertise and ability to apply it to meet specific requirements will also vary greatly. For example, it is rare that one supplier could offer an optimum solution to meet a diverse range of communication requirements ranging from traditional LANs and PBXs to wireless alternatives and high-speed LAN internetworking equipment.

It is often best to keep the number of suppliers as low possible to minimize the opportunity for buck passing. The communications industry is not yet mature, and technical problems are still relatively common. The amount of time required to solve any problems that may arise increases greatly as the number of suppliers increases. Some commentators, however, argue that breaking up "IT needs and awarding them to multiple providers . . . makes it much less expensive to switch suppliers or to bring an (outsourced) service back inhouse if a supplier proves to be disappointing" [1]. There is also a danger that excessive limiting of supplier numbers can be counterproductive in other ways. For example, in implementing such a policy, some companies issue limited tenders restricted to existing tried and trusted suppliers. While these suppliers might have been the optimum choice initially, it is unlikely that they could retain their predominant position indefinitely. Such an approach could also prevent niche suppliers with innovative, high-quality solutions from being considered. Such solutions might offer a considerable ongoing cost saving or give a company a significant competitive edge.

15.1.1 Contract Duration

The optimum contract term depends on the particular circumstances and the products or services involved. The communications industry is evolving rapidly, and a short-term contract (e.g., one to two years) allows a company to take full advantage of new products and services coming onstream. Short-term contracts also facilitate regular reviews of a company's communication requirements and allow alternative solutions that might better meet changing business needs to be more readily implemented. In addition, a supplier is less likely to become complacent or risk abusing his position if the customer can move his custom elsewhere at relatively short notice.

On the other hand, long-term contracts (e.g., 5 to 10 years) also have their strengths. Such a contract might be essential in some cases (e.g., to cover ongoing support for a privately owned communication network). In other cases, contract duration can be an important element of the overall negotiating strategy. One possible compromise is to have an initial contract period followed by one or more option periods. A typical 10-year contract might have renewal options after four and seven years. With contracts such as these, capital costs can be allocated over a guaranteed minimum period. Meanwhile, specific contract terms can be opened to renegotiation at designated intervals. Terms that might typically be renegotiated include:

- Price;
- Service levels;
- Provision of new services;
- Cessation of obsolete services;
- Geographic scope.

There is no necessity to make a final decision about contract duration at the initial tendering stage. Instead, this process could be used to evaluate one or more of these options by allowing a range of alternative responses. This could provide valuable additional information and hence facilitate a more informed choice.

15.2 External SLAs

Service level agreements (Chapter 14) are often the most important document governing the supplier-customer relationship. In the context of a communication network, the service being provided may range from the maintenance of a small PBX system to supplying an entire communication solution to a large multinational corporation. The greater the scope of the service, the greater the importance of the SLA.

15.2.1 Typical Structure

It is difficult to define a typical SLA, as they can be diverse both in their scope and in their level of detail. However, many SLAs will incorporate the following sections: a service definition, performance criteria, a list of any support structures to be provided, a list the procedures to be followed should any difficulties arise, and any additional facilities.

Finally, while SLAs can be legal documents in their own right, it is recommended that they be an attachment to a legal contract. This approach facilitates changes, by mutual agreement, without having to redraft a legal document with all the attendant time and expense implications.

15.2.2 Service Definition

A service definition can range from a comprehensive document underpinning an outsourcing contract to a brief specification covering a particular communication service. It should be as concise as possible while being unambiguous and readily understood.

Table 15.1 illustrates one possible service definition for a very basic SLA between a service provider and a business customer.

15.2.3 Performance Criteria

The selection of suitable performance parameters and their target values is the keystone of any successful SLA. In the case of external SLAs, they are often the basis of the entire contractual relationship with the supplier. It is not unusual for both the service provider and service taker to measure the same performance parameters independently. In such cases, a procedure should be agreed for dealing with any discrepancies that may arise. This can be a particular challenge where one party has a more sophisticated measuring capability than the other

Table 15.1
Service Definition for a Basic SLA

Service type	64-Kbps leased line
Lead time*	Two weeks
Tariffs	As per tariff schedule
Discounts	10% discount for 10 or more lines
Availability	99.985%
Fault index	Maximum 3 faults/line per year
Maximum downtime	Four working hours**

* Lead time is defined from date of placing order to date of acceptance of delivery by customer
** For any individual outage

and hence can produce statistics that are not readily verifiable by the other party.

The performance criteria chosen are particularly important with more complex solutions, where there can be a number of different locations each having equipment from a range of manufacturers. In such cases, there can be a danger that some performance issues are inadequately addressed leaving room for future disputes over service levels. An example might be neglecting to specify that a specific performance figure applied to all nodes in the network rather than one or more particular nodes.

15.2.4 Service Support

The type, frequency, and format of performance reports should be listed here along with a list of recipients—on both client and provider sides. It can be useful if such reports, based on the preagreed performance criteria, are supplied as a computer spreadsheet, which greatly facilitates both trend analysis and comparison with customer-generated reports. There should also be a fault-handling procedure detailing how faults should be reported to the supplier and the subsequent handling of them. The supplier should be obliged to keep the client informed of progress in fault rectification at specified intervals.

Regular formal meetings with the service provider should also be scheduled to discuss the foregoing reports and the short- and long-term objectives of each party. In addition, there should be a major customer satisfaction review at periodic intervals—say, twice yearly. If possible, this should be attended by more senior management representatives from both parties. This should ensure that there are no ongoing difficulties of which senior management, of either side, are unaware.

In some cases, the client may be open to fraud due to service or equipment capabilities outside of its direct control. Where this is the case, the supplier should be obliged to follow documented procedures to minimize the risk of such an occurrence. Procedures should also be agreed to facilitate early detection of security breaches and their subsequent rectification. In other cases, the client will have direct control over the resources that are vulnerable to attack. Here, it should be the responsibility of the service provider to highlight any security implications of the service being provided. Examples might include advice on the correct configuration of equipment or on how to implement suitable password control.

An escalation procedure should be detailed, to be followed should progress in solving any particular problem be deemed inadequate. The name and rank of both supplier and client personnel to be contacted should be listed. An escalation procedure can be progressive—the first stage of the escalation procedure might be invoked if, for example, an outage exceeded three hours, with further escalations if the outage becomes more prolonged. It can also be useful to make provision for an independent arbitrator if agreement cannot be reached internally.

Additional facilities not central to the primary service may also be provided. These can have a significant effect on the value of the overall service provided by the supplier.

Examples include:

- *Customized billing.* This might be central billing for multiple services or single currency billing for a multinational service (e.g., Syncordia manages all of BP's telephone charges with third parties and provides them with comprehensive records [2]).

- *Single point of contact.* A single telephone number might be provided for all sales and technical issues.

- *Help desks.* These are particularly useful at the early stages of outsourcing projects and for ongoing hardware and software support.

- *Dial-in remote support.* Where a vendor has communication equipment onsite, dial-in remote support allows such equipment to be interrogated and often reconfigured from a remote location. This allows faults to be diagnosed and often rectified much faster, hence greatly improving fault response times. However, the same security concerns apply here as apply to other types of remote access.

15.3 Going to Tender

Where a supplier is to be selected as a result of a formal tendering process, a *request for proposal* (RFP) or *invitation to tender* (ITT) must be issued. Depending on the particular circumstances, this may have a limited circulation or be publicized in such a manner as to encourage a broad spectrum of responses. In either case, the overall objective is to encourage high-caliber replies from suitably qualified companies. These should be in a format that allows competitive, compliant proposals to be readily identified and, subsequently, ranked in order of merit.

If this objective is to be achieved, the structure and content of the RFP needs careful consideration. If it is unnecessarily complex, it might discourage potentially suitable suppliers from applying. On the other hand, it must be sufficiently detailed that all client requirements are clearly and unambiguously specified. Additionally, the RFP must be structured in such a fashion that potential suppliers can supply all the required information with a minimum of effort. The information must also be supplied in a manner that allows easy comparison between competing bids. One possible structure is shown in Figure 15.1, and each section will now be considered in more detail.

15.3.1 Executive Overview

A brief resumé, highlighting salient points, can aid prospective tendees in deciding whether to tender and can also be useful when briefing appropriate inhouse personnel.

15.3.2 Introduction

The introduction normally commences by stating the purpose of the tender (e.g., Irish Telecommunications PLC invites potential suppliers to submit a proposal for an integrated voice and data network linking its four sites). This may be followed by a brief outline of the structure of the RFP, followed by some general clauses such as:

- A confidentiality clause to restrict the circulation of the RFP (e.g., to tendees' employees directly involved in preparing a response);
- A validity clause to specify the expected validity of any submitted proposals (e.g., for 90 days following the closing date for receipt of proposals);

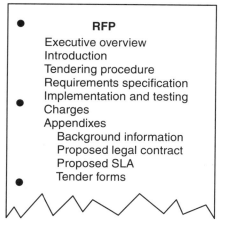

Figure 15.1 Typical request for proposal.

- A submission clause to state the conditions under which the proposal is submitted (e.g., whether modification of the initial bid will be allowed or whether the lowest bid will be accepted).

15.3.3 Tendering Procedure

The tendering procedure outlines how the overall tendering process will be carried out. Central to this is the tender time scale, which lists key activities and corresponding dates. A formal time scale (Figure 15.2) increases the efficiency of the procurement process for all parties involved. It ensures that each party is available at the prerequisite times and can avoid much time wasting. Where specific deadlines are listed, it should be clearly stated whether these are aspirational or should be strictly adhered to.

Also included in this section are instructions on how to structure the tender response. Here, the key objective is to receive all replies in a standard format that minimizes the processing required during tender evaluation. This is often achieved by designing a series of standard forms that must be completed by all respondees. Where this approach is being taken, the tendee will be referred to the appropriate appendix. Whether or not nonstandard proposals will be considered, and if so under what conditions, may also be specified. Such an option allows for the possibility of an innovative response that cannot readily be structured in the prescribed manner. The acceptability of joint bids from multiple suppliers should also be specified.

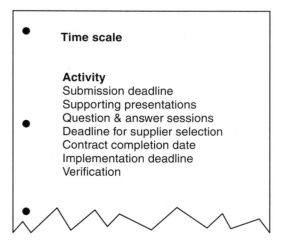

Figure 15.2 Typical tender time scale.

A circulation list will be provided listing all parties who must be provided with copies of the documentation. Many larger companies now insist that all responses be prepared using specified software packages and then emailed to a defined electronic circulation list. This practice is set to grow as the spread of the Internet and other similar services makes email almost ubiquitous. In such cases a list of approved software packages should be provided to avoid any incompatibility issues. This approach also allows limited security measures be implemented through use of the built-in encryption facilities many such packages have.

15.3.4 Requirements Specification

The requirements specification is used by potential bidders in assessing their ability to respond to the RFP. Compliance with the requirements specification is a also a key parameter in the supplier selection process. Preparation of a requirements specification has been covered in Chapter 14; however, it is worth noting that it should not be unnecessarily restrictive. There can be a risk that the requirements are specified in such a manner that only a limited number of solutions can be compliant. This could restrict the number of suppliers qualified to bid, or it might dissuade others with potentially attractive solutions from replying.

15.3.5 Implementation and Testing

Any particular requirements with regard to the project implementation are addressed here. For instance, there might be a specific period during which all equipment must be installed, configured, and tested. As an example, one large supermarket complex wanted 25 cash registers, and associated support systems, upgraded in an 18-hour window between 6 PM Saturday evening and noon on Sunday. This particular project was completed successfully, to the great relief of the vendor who faced severe performance penalties if the deadline was not met.

Acceptance testing should always be carried out before any contract is regarded as satisfied. In most cases, the main objective of acceptance testing is not, as it is often misconstrued, to confirm conformance to a standard. Manufacturers will normally supply independent certification of conformance to any required standard and, if necessary, tendees can also be asked to complete a standards-conformance form. Rather, the main objective is to ensure that all equipment interworks satisfactorily and fulfills the requirements specification. This can by no means be assured through standards compliance alone, particularly where equipment from different manufacturers is involved.

In many cases the scope of the acceptance tests will be such that they can be fully detailed in this section. In a small number of cases, detailed conformance tests will be necessary—for example, where equipment must conform to a particular subset of a standard. In such cases the detail of the particular tests required will often form part of the requirements specification.

15.3.6 Charges

Control over the manner in which a prospective supplier presents his or her charges is crucial. It allows an accurate indication of total cost be obtained and aids cost comparisons between competing vendors. It also ensures that the information is presented in a format that conforms with company accounting practices and hence facilitates any inhouse approval process. This section may, for instance, state that the charges listed must be fully inclusive and that any additional costs incurred (e.g., for equipment installation or software licenses) will be the supplier's liability. Charges will normally have to be presented in a specific format or using standard forms (see Section 15.3.7, Tender Forms). For example, it may be necessary to indicate projected expenditure over a specific period—say, five years—with a breakdown under specific headings. Discounts may also have to be calculated in a particular manner (e.g., using published tariff sheets as a baseline) and the commitment required (e.g., volume or duration based) fully listed. Where a proposed legal contract forms part of the RFP, the relevant sections on charges should also be referenced here.

Standard forms are not always the optimum means of comparing charges. One problematic area is where there is considerable diversity between the various products being offered. Another is where the procurement is for an ongoing project, which is difficult to dimension exactly. In such cases, a useful alternative is to develop one or more typical scenarios to be used as the basis of each response. One pricing scenario might be a listing of all of the requirements of a typical network node. Another might be a complete network comprising a number of such nodes.

15.3.7 Appendixes

Background Information

Company information, which may be useful or indeed necessary, in preparing a tender response is given here. This may include the geographic spread of the company, the number of employees, and turnover. Sometimes a copy of the annual report is also included.

Information of a highly technical nature may also be provided, particularly where an enhancement or replacement of an existing solution is being sought. Such information may include network diagrams, traffic measurements, and a brief description of any existing problems.

Proposed Legal Contract

It is recommended that the client drafts a standard contract to be included as part of the tender document. Such an approach ensures a level contractual playing field and makes it the responsibility of the supplier to highlight any specific concerns they might have. It also provides protection against hidden clauses in suppliers' standard contracts, which could result in unforeseen costs or other difficulties at a later date.

Standard contracts also aid in the comparison process between different suppliers and enhance client flexibility and control.

Service Level Agreement

The ongoing relationship between the supplier and the client is determined in large part by the SLA. With a well-written SLA, both parties are fully aware of their responsibilities and of the consequences of not meeting them. The SLA can also act as a barometer for measuring the ongoing supplier-client relationship.

Tender Forms

The aim of tender forms is to ensure that all tender responses are received in a standard format, minimizing the processing required during tender evaluation. This means that, for maximum efficiency, the tender evaluation criteria should be considered before these forms are designed. Such an approach allows tender forms to be designed to ensure that all key decision information is readily available. It can avoid much time spent searching through responses for the required information or having to refer multiple queries back to the tenderer. Unsuitable suppliers (e.g., those failing to meet mandatory criteria) can be quickly highlighted and the effort required to shortlist potential suppliers greatly reduced.

A set of standard tender forms is normally used, with a typical range being shown in Figure 15.3. Of these, the statement of compliance is by far the most universal found in most tender documents. It is a detailed response to each paragraph of the tender document and its effectiveness depends on how well this has been structured. For instance, if all mandatory elements of the RFP are grouped together, it is far easier to eliminate responses from unqualified suppliers.

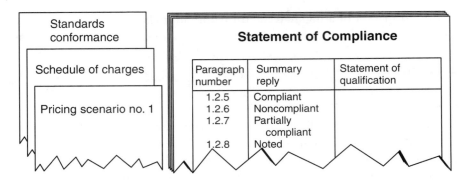

Figure 15.3 Standard tender forms.

There are three sections to a statement of compliance as shown. In section one the supplier lists each paragraph sequentially. In the summary reply section there can be one of four responses: compliant, partially compliant, noncompliant, or acknowledged—implying that an information paragraph has been read and understood. The third section, or statement of qualification, allows the tendee to give additional information in relation to the summary reply. For instance, this might explain why a proposal is only partially compliant and the implications of same.

A refinement of the tender form process becoming popular is to design the forms as software templates that must be completed by the tenderer. This makes the information readily accessible for a wide range of manipulations and further aids in the decision-making process. Where this approach is being taken, a copy of the templates will be included with the RFP. In such instances, the software to be used in preparing responses, including version numbers, should be fully specified if incompatibility issues are to be avoided.

15.4 Selecting Suppliers

The time and effort required to select a suitable supplier can vary greatly. For example, the selection of a maintenance contractor to maintain a small single-site network might take a couple of days. Alternatively, it can take a number of months to select a vendor for a large outsourcing contract.

As a general principle, it is recommended that a representative cross-section of managerial disciplines be involved in any selection process. This should, for example, avoid the risk of the optimum technical solution being chosen to the detriment all other factors.

15.4.1 Generating a Short List

The first stage in selecting a supplier is to establish general criteria that will be used to generate a short list of potential suppliers. It is not recommended that cost considerations be considered at this stage, even though they will play a critical role at a later stage. Generation of a short list will be greatly facilitated if the RFP requests mandatory responses, in a standard format, as discussed previously. Typical criteria to be used would include some or all of the following, as appropriate:

- *Compliance with mandatory RFP elements.* Any tender response failing to meet a mandatory RFP requirement should be automatically eliminated.

- *General issues.* This would include the presentation and professionalism of submitted tender and its overall compliance to the RFP.

- *Track record.* The supplier who has only recently obtained an agency for the products or services being offered might not be the optimum choice for a mission-critical project.

- *Geographical reach.* Where the RFP is for a communication solution spanning a number of locations, and ongoing support is required, it may be considered necessary for any potential supplier to have a point of presence in each location. Should this be the case, consideration should also be given to any expansion plans currently envisaged.

- *Quality assurance.* A recognized quality program (e.g., ISO 9000) is an essential supplier prerequisite for any large-scale procurement. However, such a program should not be used as an excuse for reduced vigilance, as it is by no means a guarantee that nothing can go wrong.

Other criteria needing consideration are the supplier's business strategy and product-support facilities as discussed in the following sections.

15.4.2 Supplier's Business Strategy

It can be very difficult to get an accurate impression of a supplier's business strategy, even though this will have a major impact on the service provided. At the low end of the market, some equipment suppliers compete mainly on price with little or no emphasis on customer service. Such a supplier might suit a small company with limited financial resources and a low dependence on the particular communication service. Alternatively, this supplier might be used for commodity items where there is a significant cost differential.

Other suppliers, both large and small, are there for the long haul. They take a strategic approach to their business and attempt to offer a complete service. Depending on their size, they can offer a range of services in addition to basic product or service delivery. Many also take a keen interest in their clients' business environments and can be a valuable resource in helping to form a business strategy. Such quality suppliers will charge a premium for their services and a value-for-money assessment will have to be made.

A critical evaluation of business strategy should be carried out for each potential supplier on the short list. For large-value mission-critical contracts, this should be particularly rigorous, more particularly where outsourcing contracts or multiple suppliers are involved. A useful starting point is the manner in which the potential supplier's salespeople and other agents handle queries and follow up on commitments. It can also be useful to visit a potential supplier's premises. For larger clients, it is worth investigating if any other section or department has had dealings with a particular supplier. Finally, additional information can also be obtained from reference customers.

When evaluating a potential supplier's business strategy, the particular areas that might be considered include:

- *Understanding of industry sector.* This is a major issue where the client has limited inhouse expertise and is relying on the supplier as a strategic partner. Ideally, such a prospective partner should have a proven track record in using communications technology to help businesses maximize competitive advantage.

- *Commitment to working with other suppliers.* Where multiple suppliers are involved in providing an overall communication solution, the potential supplier should be asked to document any previous experience of working in such an environment.

- *Business improvement strategy.* Evidence should be sought that the potential supplier is increasing their efficiency on an ongoing basis and hence should remain cost competitive.

- *Strategic alliances.* Many of today's suppliers are members of strategic alliances that can offer a number of advantages including access to a wider range of products or services and more POPs.

15.4.3   Product Support Facilities

As business' reliance on ICT increases, the effect of a loss of a communications service can be significant or even critical. Good support facilities are vital in minimizing the risk of such an event occurring and, should it occur, minimizing the duration of the resultant outage. This is particularly important in the case of hardware or software suppliers, who are the focus of this section. Often of equal importance is the evolutionary strategy for the products. If a substantial investment is to be made, particularly in communication equipment, every effort should be made to procure a future-proof product. It is worth noting that getting verification that a product is upgradable is of little use if the necessary hardware or software has yet to be developed and the manufacturer is concentrating on developing new improved products. Where the vendor is providing a complete service, many of the issues discussed here will be part of his or her overall responsibility.

The supplier should have a clearly defined upgrade policy for both software and hardware. This should include the frequency of upgrades and ongoing support, including spares availability, for previous issues of software/hardware. Many suppliers, including PBX vendors, only support a limited number of previous hardware or software releases, and, in effect, force users to upgrade regularly. A proposal that looks very cost effective initially can be far less attractive when the cost of these upgrades is factored in. Other suppliers charge for "new" software that must be installed to address malfunctions in the original product. This can lead to significant maintenance charges, which might not be budgeted for. To avoid these unpleasant surprises, all costs should be fully documented and agreed to in advance.

It should be the supplier's contractual responsibility to supply and fit maintenance releases of hardware and/or software free of charge for a specified period of at least one year and preferably five. The supplier should also be contractually obliged to advise the client of any known problems with its product. At least two well-known manufacturers of communication equipment do not advise their clients of any ongoing problems that arise. In many other cases, such notifications get no further than the distributor. Having spent a number of days fault-finding on a network, it is extremely frustrating to find that the problem is due to a design fault of which the manufacturer has been aware for some time. Problems are particularly likely to arise where prerelease (or other nonstandard) software/hardware is supplied to meet an urgent need.

Where support or maintenance is to be carried out inhouse, access to reliable up-to-date information and training is essential. In addition, a technical-support channel is almost mandatory in such cases, especially where software products are involved. The hours of availability of such support is also important, as can be the language. Where the product is not manufactured in an English-speaking country, it can sometimes be extremely difficult to get intelligible technical support. The supplier should also be asked to state their documentation distribution and upgrade policy. Some manufacturers do not update their documentation in line with production changes, which can make fault-finding extremely onerous. Training availability and cost should also be documented. In particular, training should be quickly available for new products or upgrades.

Choosing a Supplier

The next stage of the selection process is to carry out an in-depth evaluation of those suppliers who are short listed. At this stage, potential suppliers can be asked to clarify particular points of detail or even to make a formal presentation to support their tenders. The particular method chosen to select the successful tendee will vary from organization to organization.

One of the most common approaches is to use a weighted ranking of all factors considered important, as shown in Figure 15.4. The level of detail involved here will depend on the complexity of the requirements, the size of contract, and how critical the final outcome is to the business. For large contracts, an extensive appraisal may be required using the tender documents as the basis for the assessment. Each requirement is normally graded out of 10, as it is difficult to work to a greater level of precision. The individual result is then multiplied by a weighting factor, which indicates the relative importance of each requirement. The maximum weighted score is often chosen to be a round number (e.g., 100 or 1000) to facilitate easy comparison.

The foregoing approach normally allows the main contenders be identified but rarely identifies a clear winner. One key determinant is the manner in which overall cost is assessed. In some instances, cost containment might be the critical factor—in which case a specific threshold (minimum number of points) might be set. The most cost-effective solution might now be chosen from the list of qualifying suppliers. Alternatively, value for money might be more im-

Requirement Number	Weighting Factor	Supplier 1		- - - - - - - - - - - - Supplier N
		Raw Score	Weighted Score	
1	7	(1-10)		
2	15			
3	8			
Totals	100		100 max.	

Figure 15.4 Simple weighted ranking form.

portant, and here the number of points per unit cost is sometimes used. For example, supplier 1 might score 750 points and have a total cost of $1,125,000 giving $1,500/point. Supplier 2 might score 800 points with a total cost of $1,240,000 giving $1,550/point. In this case the first supplier would be chosen, even though she scored fewer points. Additional factors that might be considered include how well employees and vendor staff mix and the supplier's business strategy.

15.4.4 Reference Customers

Where a potential supplier isn't well known to the client or doesn't have a proven track record, it is normal practice to contact some of his or her existing customers. Ideally, these will be identified through business contacts or a trade association. Failing this, the supplier should be asked to provide such a list. Obviously, this is less than ideal as any competent supplier can be expected to prescreen those selected. In either case, a random selection should be contacted and opinions sought. If possible, the selection should include one or more customers of similar size to the client and with similar requirements. They should also be using the product sufficiently long to be able to give an authoritative assessment.

Ideally, any meeting with reference customers should be held without the supplier being present to allow searching questions be asked. Typical questions one might broach include:

- Was the agreed time scale met?
- Were user expectations fulfilled?
- Did promised savings materialize?
- How quickly were problems solved?
- Would you use the same supplier again?

Where a significant capital outlay is involved, this process should be followed up with a visit to one or more reference sites. A lot of useful information can be obtained from observing the quality of installation work and, where possible, talking to some of the operational staff. For very complex or wide-ranging projects, a pilot trial might be required before a decision to proceed is made. In the case of equipment suppliers, it may also be deemed necessary to visit the manufacturing facility, though certification to a specified quality standard such as ISO 9000 is usually sufficient.

15.5 Contractual Issues

A contract is a legal document that forms the basis of the relationship between the supplier and the client. A well-drafted contract should be based on a clear understanding of the rights and expectations of all parties involved and hence should reduce the possibility of any conflict in the future. A number of contractual issues of particular concern to communication services are highlighted in this section [3,4]. It is not an attempt to cover contract law, nor indeed all details of a typical contract, and it is recommended that alternative sources be consulted to gain a detailed insight into this area. In particular, where there are significant amounts of money involved, or the services involved are critical to the client, any proposed contract should be assessed by a qualified legal professional. An alternative, as discussed under tendering, is for the client's legal representative to draft a standard contract to accompany the initial request for proposal.

Ideally (see Section 15.3.7) the supplier will accept a standard contract prepared by the client. In some instances, however, the supplier will insist on using their own contracts. However, it is always prudent to review such contracts and, if necessary, negotiate for their modification to address particular concerns. Communication products and services are often customized to meet customer needs, and many suppliers will also facilitate reasonable changes to the standard contractual terms. For example, changes might be made to the guarantee period or to payment terms and schedule. In particular, it is good practice to ensure that the final payment is not due until acceptance testing is complete and the network is working satisfactorily for an agreed period (e.g., 12-month guarantee period).

15.5.1 General Provisions

Contracts for communication products or services normally have a section with so-called general provisions. These will include a list of the products or services being provided along with a broad outline of the scope of the contract and a list of attachments. The outline should state the contract duration and if it is to be an open tender for future work. Whether the contract is exclusive or the client is able to go elsewhere for additional products or services should also be covered. A renegotiation mechanism, or other means of incorporating changes, should also be considered to limit a client's exposure to exploitation. This is particularly important in cases such as outsourcing contracts, where a client might be effectively locked in to a supplier on a long-term basis. Any additional charges (e.g., for equipment delivery or installation) may also be listed.

15.5.2 Pricing Policy

Many standard contracts give the supplier the right to change aspects of its product or service, including pricing, without prior notice. In some instances this might be acceptable or nonnegotiable. However, where this is clearly undesirable, an attempt should be made to negotiate a minimum notice for such changes to allow possible alternative suppliers be considered. Ideally, the contract should state that all charges will remain fixed during the duration of the contract unless more favorable terms are offered. The success of this initiative will depend on such issues as the total volume of business involved, the customer-supplier relationship, and the ongoing potential for additional contracts. The payment schedule—when payments are due—should also be reviewed. Where a discount plan is available, it should, if possible, be negotiated to cover the entire basket of supplier goods and services. This makes invoice administration simpler and can be more cost effective than having a range of different schemes from a single supplier.

In some cases a client will, in effect, be locked in to a particular supplier once a contract has been negotiated. This could apply to buying a particular vendor's networking solution or to an outsourcing contract, where the cost of reverting would be excessive. In such instances, a formal mechanism for calculating price levels should be included in the contract to avoid a customer being exploited or otherwise disadvantaged. This should allow for any savings to be passed on to the client should the underlying cost to the supplier actually decrease. It should also cover any rescheduling of charges by the supplier to avoid the possibility of the client being committed to terms less favorable than those currently on offer. This is a very real risk, as the charges for communication goods and services are continually changing in response to competitive pressures.

One possibility is to have a mechanism where the supplier cannot offer the client less favorable terms than those offered to other customers with similar requirements. This clause should apply to both the initial contract term and any subsequent renewals. An example might be two customers requiring the same number of switched circuits (e.g., SMDS connections) and having similar traffic volumes and usage patterns. Other options include benchmarking against an industry average or linking charges to inflation. Should such issues arise, the client should have the option of accepting the revised terms and conditions or retaining those already contracted for.

An alternative approach for large contracts is to negotiate a fixed fee based on agreed financial targets plus a specified percentage of any additional savings achieved by the supplier. These savings might result from improving the efficiency of the communication network or the falling cost of equipment or services. Such an approach would incorporate a mechanism for calculating and itemizing costs and require access to supplier books for inspection and auditing [2].

15.5.3 Confidentiality Clause

The contract should make a provision to protect the confidentiality of information obtained from the client in addition to the standard clause protecting the confidentiality of proprietary information provided by the supplier. In this context, *information* should be defined sufficiently broadly to include any knowledge or experience that the supplier or his or her employees or agents gain from their business dealings with the client, the disclosure of which to a third party would have a material effect on the client.

Such a clause is particularly important in the case of service providers, who often require an in-depth knowledge of the client's business if they are to fully meet their needs. In other cases, such knowledge will be acquired as a result of the very close working relationship that is the hallmark of a good business partnership. A suitable confidentiality clause avoids the risk of this knowledge being misused either during the period of service provision or subsequently. For example, the client may deploy communications technology in an innovative manner to gain a competitive advantage. It might be in a supplier's interest to use the knowledge of this deployment in his or her dealings with competitors of the client unless contractually restrained from so doing.

The confidentiality clause may also need to be extended to cover legal requirements in instances where, for example, a supplier has access to information subject to data protection legislation.

15.5.4 Liability for Fraud

Where there is a risk of fraudulent use of equipment or services, there should be a clear assignment of liability for any loss so incurred. A distinction should be made between fraud, which is the clear responsibility of the client, and that due to events outside the client's control.

Issues that might need to be addressed include:

- The failure of the supplier to meet his or her obligations to ensure system security as specified under the service support section of the attached SLA. For example, many instances of toll fraud—where hackers make free calls using a company's telephone lines—are made possible by the supplier's failure to correctly configure a PBX.

- Fraud made possible by an inherent weakness in a product or service that is impossible to eliminate. While such weaknesses are increasingly rare, they will always continue to exist. One example is the cloning of analog cellular phones for toll fraud.

15.5.5 Software Licenses

Software licenses do not come cheap and can be a significant cost for many communication networks (some network management software can cost hundreds of thousands of dollars). To protect this investment, the licensing conditions should allow the transfer of licenses to a new site. They should also allow license transfer to a third party service provider to allow the client outsource some of its communication requirements in the future, should it decide to do so. This section should also protect the client against third-party claims that the software infringes a copyright or other right and should require the supplier to provide redress.

Where a supplier is contracted to develop software for a particular application, the intellectual property rights should be vested in the client unless a suitable discount that recognizes the ongoing value of the software to the supplier is given. The supplier should also be obliged to furnish the client with all source code supported with appropriate documentation. Failure to address this issue properly could leave the client open to exploitation, as the client could become dependent on a single supplier for the ongoing development and support of a critical application. The client would be in an even more unenviable position should the supplier cease trading.

15.5.6 Multiple Suppliers

Many companies use multiple suppliers. Where a lead contractor has been appointed to a particular site, the contract should outline the responsibilities of this contractor, particularly in relation to problem ownership and resolution. The responsibilities of subsidiary contractors should also be documented if appropriate.

Where a single contractor is used, the contract should specify any particular conditions required regarding the use of subcontractors or even their prohibition.

15.5.7 Exit Mechanism

Even the best of business partners may not be at their most cooperative when told that their contract is not being renewed or is even being canceled prematurely. For this reason, many service contracts include a formal exit mechanism. This states the conditions under which either party may terminate the contract and the rights and obligations of both parties in each case. A normal termination will usually occur at the end of a fixed term when an option to renew is not taken up by either party. Typical criteria for abnormal termination include a major default by the supplier or severe financial difficulties in either case. Other criteria may be included to cater for specific concerns.

There should be a clear distinction between a normal and abnormal termination and the procedures to be followed in each instance should be specified. For example, the contract might stipulate that the client must give written notice of a material breach of performance criteria specified in an associated SLA. An agreed period (typically 30 days) would then be allowed for the service provider to rectify the breach, following which, the client would have the right to terminate the agreement. Any financial penalties arising from an abnormal termination should also be specified. This is particularly important in the case of communication outsourcing, where there may be substantial costs involved in changing service providers or there might not even be an alternative service provider. In the latter case, it would be prudent for the client to insist on an appropriate penalty clause to cover the costs of reverting to a private network solution. Such clauses should make provision (e.g., liquidated damages) to ensure that resources will be available to meet any penalties incurred.

The means by which service is maintained during the changeover period may also need to be specified. In many instances it is essential that the service provider continues to provide full support while the changeover is in progress. The level of support required will vary from case to case but might range from a continuation of the contracted service for a limited period (typically 30 to 60 days) to the temporary transfer of key service provider staff to the client. The manner in which any nonstandard support required during the changeover will be charged should also be documented. Other issues, such as the transfer of any remaining assets or other property, should also be addressed.

15.5.8 Attachments to the Contract

Supporting documents are often attached as appendixes to a contract. They can help clarify the intent of the contract and are useful should disputes arise at a later date. They can include such items as suppliers' specification sheets and advertising copy. Contracts awarded as the result of a tendering process often have the original RFP, along with the tendee's reply, also attached.

When the supplier is a service provider, the contract should be linked to the SLA in such a manner that changes to it, by mutual consent, are not precluded. This allows sufficient flexibility to develop the service in light of experience (e.g., the performance criteria may need fine tuning). However, care should be taken that there isn't a conflict between the legal contract and the linked SLA to avoid any disputes at a future date. This would be particularly important in the case of an abnormal contract termination where litigation might result. The contract should also give a legal basis for any penalty clauses (e.g., rebates) in the SLA that may need to be invoked if the obligations of the service provider are not met. It might also require lodgment of a performance bond to ensure that any money due will be available should the supplier be unwilling or unable to pay the appropriate penalties.

Some supplier contracts are awarded on the strength of the expertise of a small number of employees. In this instance, a list of supplier personnel filling key positions should be identified by name in a formal attachment to the contract. The mechanism through which new staff can be assigned to these posts should also be documented.

15.5.9 Outsourcing Contracts

Because the level of risk is so great, the caliber and composite of the contract management team is critical when negotiating outsourcing contracts. The team should include people with contract management skills, legal expertise, and technical experts who fully understand the requirements specification.

The contract should explicitly address any particular concerns identified at the risk-assessment stage of the project, and any associated penalty clauses should fully reflect the effect on the business of the service provider failing to meet his or her obligations. For example, the contract might oblige the service provider to ensure that the solution he or she provides evolves in line with best practice in the client's industry sector as established by means of specific guidelines. Which party pays for such upgrades should also be covered. A mechanism for fine tuning the contract, in light of experience, might also be considered. Some outsourcing projects are very complex and an amount of flexibility can benefit all parties.

Where existing assets are being transferred to the service provider, they should be assigned an agreed valuation, and any existing contractual obligations should be addressed. Typical areas to be considered would include any leases on existing equipment and the terms and conditions of existing software licenses. The date and means of service changeover should also be detailed to ensure that the business is not unnecessary disrupted as work proceeds.

The optimum term for an outsourcing contract depends on the feasibility of changing service providers or reverting to a private solution. A client's request for additional features on a VPN, for example, is much more likely to be quickly agreed, on the most favorable terms, if the vendor knows that he or she risks losing the entire contract within a relatively short time scale. This is backed up by recent research, which indicated that 83% of short-term contracts (defined as less than four years in duration) were successful, compared with 40% of long-term contracts (seven or more years) [5]. However, in some cases, the client is effectively locked into the service provider once the decision to outsource is implemented. In such cases, the contract duration should be negotiable depending on what incentives are available from the service provider. Where any long-term contracts are being negotiated, provision for unforeseen changes in the client's business environment should be considered (e.g., geographic contraction due to significant loss of market share).

The outsourcing market is becoming much more competitive thanks to regulatory initiatives such as open network architecture in the United States and open network provision in Europe. As a result of competition in their core markets, telephone companies are becoming much more aggressive players in the outsourcing arena as they seek to diversify. This means that some players may offer strong incentives to enter long-term contracts as the battle for market share intensifies. These may take the form of loyalty bonuses, guaranteed migration policy to the most up-to-date technology or customized solutions at no extra cost.

15.6 Ongoing Supplier Management

Good supplier management is designed to create a partnership between the supplier and the client so as to maximize the value that each derives from the relationship. By adopting this approach, from the procurement process onwards, many potential difficulties can be avoided. Conversely, many companies concentrate on getting a product or service at the lowest possible price without any consideration of either their own or the supplier's long-term interests. This approach can encourage suppliers to cut their margins to unrealistic levels to secure the contract. It generally results in poor-quality service and encourages the supplier to indulge in opportunistic behavior at the first available opportunity. This might include such activities as extortionate pricing for additional equipment and upgrades or favoring more profitable customers should a conflict arise.

A well-managed relationship, on the other hand, should result in the supplier having a vested interest in maintaining a good working relationship and doing everything in his or her power to ensure that everything is running smoothly. One means of furthering this objective is to use a formal SLA, where the expectations of all parties are fully documented. Using SLAs also means that should any difficulties arise, they can be dealt with in a constructive fashion using a prearranged mechanism. This should minimize the risk of the relationship deteriorating to a level where it would be impossible to restore good working relations in the future.

15.6.1 Managing Diverse Suppliers

Where the client has significant inhouse expertise and is managing its own network, the simplest approach is to have individual SLAs with each supplier. The difficulty here is that such agreements do not normally provide incentives for the different suppliers to cooperate. This can result in significant inhouse resources being tied up in solving disputes where each supplier blames the other should any interoperability issues arise. Ideally, an umbrella agreement should be brokered between the various suppliers with a formal mechanism for solving any difficulties that may arise.

Where multiple suppliers are providing a complete communication solution particular care is needed, especially if the inhouse expertise is limited. In such cases, the possibility of negotiating a lead supplier, with overall responsibility for the complete service, should be considered. In any case, the overall objective should be to put together a package that delivers a seamless service to the business. If possible, suppliers who offer complementary products and who are committed to working well together should be chosen. This should facilitate the drafting of complementary SLAs with the active involvement of all parties. If well drafted, these agreements should avoid the possibility of excessive buck passing and encourage maximum cooperation.

In the case of a company with an international private network, the difficulties involved in managing multiple suppliers become more intense. Such a network may involve a number of different carriers and may even use several carriers' networks to cover one network segment. In the event of inadequate performance or indeed a total failure, it can be difficult both to identify the cause of the problem and then to force the appropriate carrier to accept responsibility for it—and this does not even mention any language and time-zone difficulties that may arise. One possible solution is to outsource to an international service provider offering a single point of contact.

References

[1]   Lacity, Mary C., Leslie P. Willcocks, and David F. Feeney, "IT Outsourcing: Maximize Flexibility and Control," *Harvard Business Review,* May-June 1995.

[2]   Cross, John, "IT Outsourcing: British Petroleum's Competitive Approach," *Harvard Business Review,* May-June 1995.

[3] Graham, Rory, "Outsourcing IT: Avoiding the Legal Pitfalls," Data Pro CD-ROM Service, Nov. 1995.

[4] O'Donnell, Brian, "How to be Outsourced and Happy," Data Pro CD-ROM Service, based on a presentation given at *Networks '93,* Birmingham, U.K., June 29–July 1, 1993. Copyright Newburn Consulting 1993.

[5] Lacity, Mary C., Leslie P. Willcocks, and David F. Feeney, "Letters to the Editor," *Harvard Business Review,* July-August 1995.

16

Tariffs

16.1 Introduction

We have already seen some details about tariffs in previous chapters. In this chapter we will look at:

- Components of a tariff;
- Tariff comparisons and break-even analyses;
- Sources of tariff information;
- Tariff trends.

Our prime focus will be to show the reader how to choose the best combination of service, service provider, and tariff plan to minimize communication costs. We also indicate what to expect in a tariff schedule.

Possibly the biggest problem with a competitive environment is the confusion it tends to create over which service provider to choose from. While the overall effect of competition is to lower the cost of basic services, one should not forget that lowering tariffs is never the prime objective of a service provider. The service provider's prime objective will undoubtedly be maximizing profits. Lowering of tariffs is a long-term means of increasing profits through increased market share. Other short- or long-term means of increasing profits include:

- Making tariffs lower on the agenda by highlighting quality of service, support, security, global networks, and latest technology;

- Confusing the tariff issue by offering a very good deal on some portion of a service such as free calls at certain times, cheaper calls to your most frequently dialed numbers, and free calls included in your rental;

- Locking a business into a solution, such as a VPN service, that will require a heavy investment on the part of the business to change;

- Providing added value for which a premium can be charged;

- Charging high rates for services that have no competition.

These techniques make it difficult to choose between service providers. First, price is not the only issue, and second, price comparisons are not always straightforward.

16.2 Comparison of Options

A Tarifica survey of large U.S. communication users shows that most rate quality and customer service as more important than price [1]. Price is thus only one of many factors to be considered in choosing between options. When this is the case, it is common to make comparisons on the basis of a weighted sum of scores for each factor, as shown in the example in Table 16.1.

An alternative approach is to use costs instead of scores, as shown in Table 16.2. For example, option A has an availability of 99.95% (to be written into an SLA). Over a three-year period (750 working days), the total expected outage time is thus 0.4 days or 3 hours. This is expected to damage the business by $5,000. Option B has an availability of 99.9%, and C of 99.99%.

Table 16.1
Comparison of Options Using Weighted Scores

Plan	Three-Year Cost (Capital + Operating Costs)	Price Score	Availability	Transaction Response Times	Ease of Upgrades/ Degree of Lock-in	Total Score
Score out of		100	40	20	30	190
Option A	$120,000	83	35	20	20	158
Option B	$100,000	100	33	15	20	168
Option C	$140,000	71	38	20	25	154

Table 16.2
Comparison of Options Using Costs

Plan	Capital and Operating Costs	Availability	Transaction Response Times	Ease of Upgrades	Total
Option A	$120,000	5,000	0	20,000	145,000
Option B	$100,000	10,000	10,000	20,000	140,000
Option C	$140,000	0	0	10,000	150,000

This method has the evident advantage that everything is converted into a common unit (money) with which everybody is familiar.

It is not always easy to assign a monetary value to a particular attribute, but the exercise is likely to lead to more realistic comparisons than weighted scores. It is normal only to make such detailed comparisons after certain options have been ruled out because they do not meet minimum performance criteria.

16.3 Components of a Tariff

Charges for communication services or facilities are made up of one or more of the following components:

- *Installation or connection:* A one-time charge when you sign up for a service;
- *Rental:* A flat charge levied periodically (e.g., monthly);
- *Usage charges:* Charges based on connect time and/or volume of data.

There is normally a charge for the basic service plus additional charges for optional features and facilities. For example, a telephone line will generally attract additional rental and installation charges if you want supplementary services such as conference calling or call waiting.

Usage charges are often the most complex. They can depend on one or more of the following:

- Duration of the connection;
- Volume of information transmitted;

- Distance;
- Destination (on international calls, some countries impose high completion fees);
- Time of day;
- Type of call (e.g., in ISDN, data calls are often more expensive than voice);
- Requested bandwidth (this is a relatively new concept that already applies on dialup to some packet-switched networks but may become important for switched frame relay and ATM);
- Volume discounts.

On the following pages we will outline the components of the tariffs for various communication services, beginning with the PSTN.

16.3.1 PSTN Tariffs

PSTN tariffs consist of connection, rental, and usage charges. The usage charges depend on call duration, distance/destination, and time of day.

Connection and rental charges are small in comparison to usage charges for a typical business. The European average is about US$130 for connection and US$130 per annual rental [2]. The average U.S. connection fees are similar, while rental is about US$400 per year for a business line but as low as US$100 per year for a residential line. Many European telephone companies admit that their rental charges are subsidized by the usage charges. It is interesting to note that rental charges are higher in countries such as the United Kingdom, which have competition for local telephone service.

Pulse Metering and Per-Second Billing

Usage charges normally begin when the called party answers the call. In some networks, however, charging starts earlier. For example, in the United States, some cellular operators begin charging as soon as you make contact with the network. In Europe, usage charges have traditionally been based on a system called *pulse metering,* where an electrical pulse is used to advance a mechanical meter in the local telephone exchange. Each meter pulse has the same money value, but there are more pulses per minute for long-distance or international calls. In the United States, usage charges are *per minute,* with different rates based on the time of day and distance. In some instances, the charge periods are six seconds rather than one minute. The advent of computer-controlled exchanges means that billing systems can be made more flexible. The concept of periodic charges has not disappeared, however.

The alternative to pulse metering or per-minute billing is often called *per-second billing*. The differences between the two methods is illustrated in Figure 16.1. In this example, the average cost per minute is the same but the actual cost of a call will depend on the exact time the call was terminated.

As can be seen from Figure 16.2, per-second billing is not an advantage in itself. It really depends on what happens at the start of the call. There are three basic alternatives, as shown in Figure 16.2. In case A, there is zero cost for the call setup; in case B, there is a small charge for call setup (say, six cents); and in case C, there is a charge of, say, 45 cents for call setup plus an initial period of talk time (one to three minutes). Case C is the most common method used today.

Applications such as credit card validation or email, which require a large number of very-short-duration calls, have the most to gain from a move to per-second billing, provided the minimum charge for a call is significantly less than the charge for a single meter pulse under the old regime.

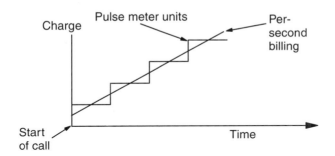

Figure 16.1 Meter units versus per-second billing.

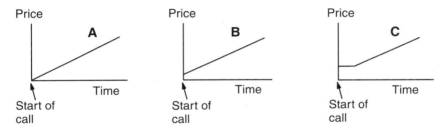

Figure 16.2 Alternative approaches to per-second billing.

Distance-Related Charge Bands

The relationship between usage charges and distance is a step function as shown in Figure 16.3.

The charge does not depend on the exact distance between the two callers, but on the distance between certain telephone exchanges. Some countries, such as the United Kingdom, use a system of *charge groups* and an associated charging matrix that defines the correct rate between any two charge groups, as shown in Figure 16.4. In this case, calls within a charge group and calls to an adjacent charge group are charged at the local 1 (L1) rate while calls further afield are charged at the higher local 2 (L2) rate.

Discount Schemes

In an effort to maintain or increase the number of customers in a competitive market, phone companies have come up with a range of discount schemes:

- Volume discounts for those who make a large number of calls;
- Free calls included in the rental;
- Discounts for calling within a VPN;

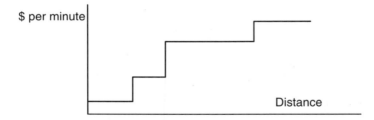

Figure 16.3 The relationship between call charges and distance.

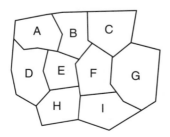

	A	B	C	D	E	F	G	H	I
A	L1								
B	L1	L1							
C	L2	L1	L1						
D	L1	L2	L2	L1					
E	L1	L1	L2	L1	L1				
F	L2	L1	L1	L2	L1	L1			
G	L2	L2	L1	L2	L2	L1	L1		
H	L2	L2	L2	L1	L1	L2	L2	L1	
I	L2	L2	L2	L2	L2	L1	L1	L1	L1

Figure 16.4 A set of charge areas or charge groups with an associated charge matrix.

- Discounts for frequently dialed numbers (e.g., "Friends and Family");
- Free off-peak calls;
- Free calls on Friday (e.g., Sprint in early 1996).

These schemes do reduce the phone bill, but they also make cost comparisons among service providers more difficult. One point to watch with volume discounts is that the discounts often apply only to calls over and above the magic volume where the discount comes into effect. Thus, a 20% discount may not be as attractive as it first appears.

PBX Trunking

In the United States, where service providers are distinguished as either local (LECs) or long distance (IXCs), the owner of a large PBX will find it economical to connect directly to the IXC exchange to get discounted long-distance calls. He will also connect to the LEC (the normal way to connect a small PBX to the PSTN) and use this connection for local calls and as an alternative path for long distance (to be used only when all the trunks to the IXC are busy).

The option exists in many countries to connect a PBX to the PSTN using an E1 or T1 digital trunk (see Section 3.2.6). Some network operators actually charge more for this service than for the equivalent number of PSTN lines (30 in the case of E1 and 24 in the case of T1), while others charge less (both for connection and rental). Call charges are nominally the same, but if you need 30 trunks you will more than likely have sufficient traffic to qualify for a volume discount.

Even if the phone company charges more per "line" for E1 or T1 PSTN trunks, it must be kept in mind that a single T1 port for a PBX costs a lot less than 24 analog ports.

16.3.2 Cellular

Cellular tariffs differ from PSTN tariffs in a number of ways:

- You pay for *air time* even for incoming calls in many places, notably in the United States.
- Roaming charges apply when you use your cellular phone on another network (see Section 9.3.1).
- Connection fees are generally lower than for the PSTN.

- Rental is generally higher, though this is sometimes offset by the inclusion of free call minutes in the rental. There are some notable exceptions, such as in Sweden, where zero-rental options exist. (This zero-rental option is accompanied by higher peak-time call charges.)

- The coverage areas for local cellular calls is often larger than for the PSTN. (Typically 50 miles versus 12 miles, respectively, in the U.S.). This can have the effect that during free calling periods (off peak), the cellular service is cheaper for certain calls.

- Short messages are charged either as a flat-rate rental or per message. In many cases they are paid for by the recipient.

Some cellular operators have different call charges depending on where you are calling *from*. For example, the PCN operators in the United Kingdom have a special low tariff when you make calls *from* your home area. Conversely, some cellular operators charge higher when you are calling from highly congested areas, such as city centers.

16.3.3 Paging

Pagers used in business usually follow the *owner pays* model with a rental for the pager that covers the cost of some or all of the messages received. In this case, the caller makes a local call to the message handler.

Consider this example: SkyTel provides two-way paging services and charges a US$399 initial fee to buy the pager and US$24.95 per month, which includes a hundred 80-character messages per month. Additional messages cost US$0.25 per message (1996 prices).

Some operators have introduced the *caller pays* model aimed at the residential market. In this case, the owner simply buys the pager and does not pay any rental. The person who wishes to send the page message makes a premium rate call (900 service in the U.S.) and either speaks the message to a voice-mail system or an operator, or keys in digits on a DTMF phone.

16.3.4 ISDN

ISDN tariffs are structured in a similar fashion to PSTN. The connection fee and line rental for a basic-rate ISDN line is generally double that of a PSTN line on the basis that it gives you the functionality of two PSTN lines and more. Most network operators charge PSTN rates for voice calls, but many charge more for data calls.

ISDN brings with it many supplementary services and many of these attract additional connection and rental charges. Additional telephone numbers are a good example of a facility that costs extra.

Local Access Line—Distance Charges

In the United States, some carriers charge extra for ISDN if you have a long local loop (greater than three miles). These extra charges take the form of a higher rental and/or higher installation charges.

16.3.5 Leased Lines

Leased-line (*private-line* in the U.S.) tariffs are generally quite simple because they include only installation and rental charges. There are no usage charges for duration or volume of data.

The installation charge is normally dependent simply on the bit rate of the line. The rental will be a function of:

- Distance;
- Bit rate;
- Service guarantees.

Effect of Distance

Typically the cost per mile of a leased line is much higher for short-distance lines than for long distance. Figure 16.5 illustrates this relationship for a 64-Kbps leased line in France.

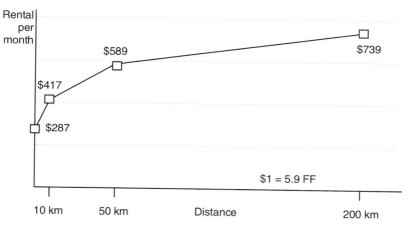

Figure 16.5 Monthly rental for a 64-Kbps leased line from France Telecom 1997.

The general shape of this cost curve is the same in most countries, but there can be large differences in the actual rates. In Europe, the monthly rental for similar leased lines can vary by a factor of four from one country to another.

Relationship Between Price and Bit Rate

There is a nonlinear relationship between bit rate and rental. The price of bandwidth decreases as the bit rate of the link increases. Table 16.3 shows the actual prices for the various bit rates in France, while Figure 16.6 shows the relative cost of bandwidth at the different bit rates. The most extreme case is a short

Table 16.3
Various Monthly Rental Charges for Leased Lines in France (in US$, 1997)

Bit Rate in Kbps	64	128	256	512	1,024	2,048	34,000
0 km	287	534	714	874	936	1,022	4,860
10 km	417	763	1,327	1,753	1,949	1,949	11,975
50 km	587	1,102	2,025	3,441	3,542	3,542	37,127
200 km	739	1,381	2,610	5,322	6,593	6,593	75,898

Source: France Telecom.

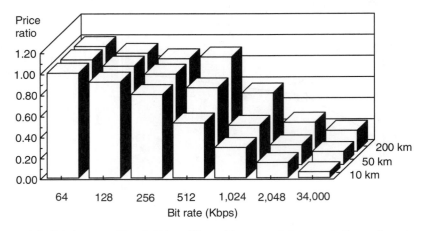

Figure 16.6 Relative cost of bandwidth at different bit rates and distances in France. In each case, the cost of bandwidth is compared relative to a 64-Kbps link of the same length.

34-Mbps line. The cost of bandwidth is only 5.6% that of a single 64-Kbps link of the same length (10 km).

Comparison of Connection Options

Figure 16.7 shows an interesting comparison between two alternative means of connecting two branch offices to a headquarters. Option B is 25% less expensive than option A, a result which strikes many people as odd until they examine the figures. The main reason for the disparity is that above 50 km, the rental is not very dependent on distance. A further advantage with option B is that the communication equipment in the middle branch office will also be cheaper.

Discounts for Long-Term Contracts

In most cases the monthly rental is lower if you enter into a long-term contract of three to five years. For example, British Telecom (BT) in the United Kingdom offer 5%, 10%, and 15% discounts on 64-Kbps national leased lines for three-, four-, and five-year contracts [2]. As already noted, five years is a long time in the telecommunications industry, and long-term contracts should be entered into with caution. In particular, you need to know the arrangements if you need to break the contract, either completely (in the case where you move to another supplier) or by changing the bandwidth of the line or by moving to a switched solution such as frame relay or ATM.

International Leased Lines

International leased lines are generally leased as two half circuits. For example, a line from the United Kingdom to the United States will consist of a half circuit leased from a U.K. operator and a half circuit leased from a U.S. carrier. The U.K. charge will include everything from NTU in the customer's premises to an imaginary point in the middle of the Atlantic ocean. The U.S. charge might

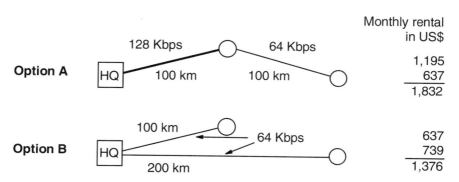

Figure 16.7 Comparison of rental for two methods of linking branch offices to headquarters.

come in two parts, one from the long-distance carrier (e.g., AT&T, MCI, or Sprint) and the other from the LEC. Many operators now provide you with one-stop shopping with a single charge for the entire connection.

16.3.6 Packet Switching

Tariffs for X.25 packet-switched networks can be quite complex, mainly because the usage charges are often based both on connect time and volume of data transferred. It is thus difficult for the average user to have a feel for how much a particular call is costing.

Fixed Charges

The rental on a direct connection will depend on the bandwidth of the connection and sometimes on the distance between your premises and the nearest network node. Distance-independent rental is common in Europe when the X.25 connection is rented from the original monopoly provider. Line rental is, however, much more expensive than for a PSTN or ISDN line, the average rental being about $700 per month for a 64-Kbps connection. Additional rental charges are levied depending on the number of virtual channels you rent and depending on whether you want facilities such as closed user groups.

Dialup users pay a small monthly rental (as low as $4 per month in some countries) but incur higher usage charges.

Usage Charges

Usage charges for X.25 calls are based on duration of the call or volume of data transmitted or on a combination of both. In most cases, there are considerably fewer distance-related charge bands than exist for the PSTN. For example, one rate exists for calls anywhere within the same country and three or four international charge bands. In contrast, few PSTN operators have less than seven international charge bands.

The volume charges are based on the number of segments transmitted and received by the caller. A segment is 64 octets of data. An octet is eight bits and is thus equivalent to one byte (or one character typed at the keyboard). Many X.25 tariffs are expressed as the cost per kilobyte rather than the cost per segment, thus making them a little easier to visualize. One kilobyte (1,024 bytes) is equal to 16 segments and is approximately equal to the amount of data required to fill the screen of a text terminal when 50% of the screen consists of blank spaces.

One aspect of per-segment billing that has resulted in high charges for the unsuspecting is that short packets containing as little as one octet of data are counted as one segment for billing purposes (Figure 16.8). This happens when, for example, you strike the page-down key to scroll a screen of data. It also happens in a less obvious way when a system is configured for remote echo. This problem can be minimized by carefully configuring the customer premises equipment.

It can be quite difficult to estimate the volume charges for a particular call. To simplify things a little, many network operators now only charge a duration charge for calls within their national network. This makes direct comparison with the PSTN or ISDN very simple. Volume charges are still the norm, however, on international calls.

Some network operators charge a minimum tariff per call. This increases the price of short duration calls such as those used in credit card validation.

Sample Application and Cost Calculation

An Irish company needs to consult a database in Switzerland that has both PSTN access and X.25 packet-switch access. What are the cost differences between making the connection via the PSTN and via the X.25 network assuming the search takes three minutes and the total data transferred amounts to 10 KB (about three typed pages or nine computer screens)?

The total number of segments is $\approx 10,000 \div 64 = 157$ for the data retrieved from the database. To this we must add the number of segments of data sent to the database. In this direction, all of the segments will be underutilized. For example, if you choose a menu option by typing a number, this data must be sent to the database immediately so one segment is sent containing the single digit typed. Assume that 20 segments are sent to the database. This brings the total number of segments to 177. The costs are calculated in Table 16.4 using 1997 tariffs.

In this example it is important to note that the quantity of data transferred bears little relation to the time spent on the connection. At 9,600 bps, it would

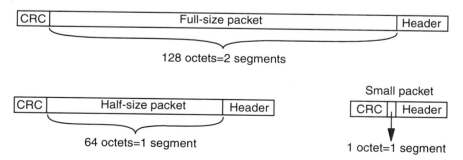

Figure 16.8 Packets are converted into segments for charging.

Table 16.4
PSTN Charges Compared to X.25

Network	$ per Minute	$ per Segment	Number of Segments	Duration Charge	Volume Charge	Total
PSTN	$0.58	$0.00	N/A	$1.74	$0.00	$1.74
X.25	$0.10	$0.002	177	$0.30	$0.35	$1.65

take approximately 10 seconds to transfer 10 KB. In this case, however, most of the three minutes is spent reading the screen, keying in search words, and waiting for results from the far end. This is where X.25 tariffs win out over PSTN tariffs. If the objective was to download a 10-KB file, the PSTN would win.

Recently, on some routes, such as the Europe–United States routes, X.25 tariffs have fallen so far behind their PSTN counterparts that X.25 can be more expensive even for the database search just described.

16.3.7 Frame Relay

Most carriers at the time of writing, only offer frame relay as a PVC service. This limits the complexity of tariffs to some extent. Many frame-relay services have a monthly rental, which is distance and bandwidth dependent but without a usage element. This is simpler for the telephone company and is attractive for customers who make heavy use of their frame-relay services.

The distance dependence in national frame relay is often limited to the access link from the customer's premises to the nearest public frame-relay node (i.e., distances between public nodes do not count). There will, however, be higher charges for interstate or international frame-relay connections.

The bandwidth dependence will be related to your port speed (the bandwidth of your access link into the network). It may also depend on the CIR and/or EIR for each PVC (see Section 7.6.2). In cases where the customers do not specify CIR, there is simply a fixed rental for each PCV.

Some frame-relay operators charge for the volume of data transmitted.

16.3.8 Internet

In most cases Internet access is charged for as a flat monthly rental with no usage charges for connect time. This is typically $10 to $20 a month for a single user access via the PSTN using a modem. Larger businesses will typically want to connect a LAN to the Internet, thus giving the possibility of everyone in the organization having access to the Internet from his or her desktop. This type of connection can be achieved using ISDN, frame relay, leased line, SMDS, or ATM. Once again, the charges do not generally depend on usage, only on the bandwidth of the access connection. They are, however, more expensive than single user accounts. For example, a 64-Kbps ISDN connection might cost as much as $300 per month on top of any rental or usage charges for the ISDN line.

In most cases, the Internet service provider will not be the same as the phone company, and you will have to pay the appropriate charges for whatever type of connection you make to the ISP, be it leased line or dialup.

16.3.9 SMDS [3]

SMDS tariffs vary in complexity. BT, for example, simply charge an installation fee and a fixed rental per site, which is related to the bandwidth of the connection at that site. For example, a 2-Mbps connection cost US$2,000 per month in early 1996. This price includes access charges and usage charges. In Germany, by contrast, the access and usage charges are separate. Access is charged at leased-line rates—this means that SMDS becomes more expensive the further you happen to be from an SMDS node. It also means that the tariff takes a quantum leap if you move from a 1.5-Mbps connection to a 4-Mbps connection. This is because the 4-Mbps access line must be provided over a 34-Mbps leased line, whereas the access line required for the 1.5-Mbps connection is only 2 Mbps.

The $2,000 per month compares well with leased-line charges in the United Kingdom. For example, a 200-km 2-Mbps leased line costs $5,000. Two SMDS connections would serve the same function for LAN interconnection saving $1,000. If three sites 200 km apart required connection, SMDS would cost $6,000, while two leased lines would cost $10,000. It should be kept in mind, however, that a 2-Mbps SMDS connection only gives 1.5 Mbps to the user.

16.3.10 ATM

ATM tariffs, where they exist, are generally related to distance and bandwidth with variations depending on the class of service used [4]. ATM services are, however, in their infancy and SVC services are not available yet. It is possible that usage charges will become more common with the eventual introduction of SVC services.

16.4 Detailed Billing

Detailed billing is available for most communication services in many parts of the world. The detailed bill is often delivered in paper format, which tends to render it useless for large organizations. Many service providers can offer billing details on tape, diskette, or via email (EDI), and this can be used for some serious analysis of traffic levels and calling patterns. Alternatively, equivalent information can be obtained from customer premises equipment such as a PBX, but you must bear in mind that some lines, such as those used for fax machines and modems, will not be measured if they do not actually pass through the PBX.

In many cases it is possible for the service provider to provide you with much more than simple billing information. Modern communication equipment has the ability to generate statistics on utilization, traffic levels, alarm indications, dropped packets, and other fault statistics. This is true even for some leased-line services.

Most users would have little use for all this information on an ongoing basis, and service providers would have even less inclination to provide it to every customer, mainly because of the increased printing and distribution costs but perhaps also through fear that it could be used against them. Some of this additional information can, however, be very useful. Periodic traffic information can be used to determine whether you have over- or undersubscribed to a service (ordered too much or too little bandwidth). Billing information can be used to allocate costs to business units. Fault statistics are an invaluable tool for monitoring an SLA. The best time to specify the statistics you may require from a system is while you are negotiating the initial contract.

16.5 Tariff Comparisons

Tariff comparisons are seldom simple. Seldom do two competing services charge the same way for every item. Differences in charge bands, differences in peak/off-peak times, and free calls or services included in the rental all combine to make comparisons difficult. Two methods of making comparisons are given next.

16.5.1 Basket of Calls

This is a straightforward method where a typical calling pattern is assumed (previous bills are a good place to get this information) and priced according to the different tariffs being compared. It is best to make the comparisons for a number of scenarios involving different traffic levels so that you get a better feel for what will happen if your calling patterns change. This is particularly important for a new business with no historical calling data. A computer model will be very helpful if you wish to analyze a large number of scenarios.

The method can also be used in RFQs, where you request a number of service providers to price out a variety of "baskets."

Tariff specialists such as those listed in Section 16.7 often publish comparisons between providers based on baskets of calls for typical businesses. These are only a first cut at a comparison. The same specialists will also be able to help with detailed comparisons based on more specific calling patterns.

16.5.2 Break-Even Analysis

A break-even analysis provides an indication of the level of traffic (usage) where one communication service or tariff becomes more economical than another. It is frequently used to compare tariffs with different structures. Examples include:

- Comparing the flat rate tariff of a leased line with the per-minute (or per-second) tariffs of the PSTN or ISDN;
- Comparing per-minute charges of the PSTN with volume-based tariffs of a packet-switched network;
- Comparing two calling plans with different amounts of free minutes.

ISDN Compared to Leased Line

In many parts of the world ISDN tariffs are based on rental plus usage (measured in minutes or seconds), while leased-line tariffs are based on rental alone. Many applications can use either ISDN or leased lines, but because the tariffs are structured differently, a break-even analysis is a good way to make comparisons. The break-even point will be the number of hours usage per month that makes both services cost the same. If usage is above this level then a leased line should be used, if usage is below then ISDN should be used. The break even-point is easier to visualize if it is expressed in hours per day rather than hours per month. The figure will work out differently depending on whether the link is required five, six, or seven days a week and also depending on whether it will be used during off-peak periods of the day.

If having made a break-even analysis you discover that your traffic levels are expected to be close to the break-even point, it might be prudent to start with a usage-based tariff and monitor the first few bills. Your bills will act as a measurement of your traffic. Based on the information gathered, you will be able to make an assessment as to whether a flat-rate tariff would be more beneficial. A second reason for taking this approach is that the installation cost for ISDN is lower than that for a leased line.

If you start with the flat-rate tariff then you will have to make special arrangements to make traffic measurements.

Calling Plans Compared

Southwestern Bell in Kansas and Missouri had three calling plans for ISDN in March 1996, as shown in Table 16.5.

The three tariffs are plotted on a graph of cost versus hours of usage in Figure 16.9. The break-even point between the 10-hour and the 80-hour plans is 17.5 hours per month or 48 minutes per day (based on 22 working days in

Table 16.5
ISDN Calling Plans

Plan	Monthly Rental	Free Hours	Cost of Calls per Minute	Cost per Hour
10-hour plan	$57.3	10	4c	$2.40
80-hour plan	$75.3	80	2c	$1.20
Flat rate	$104.3	Unlimited	0c	$0.00

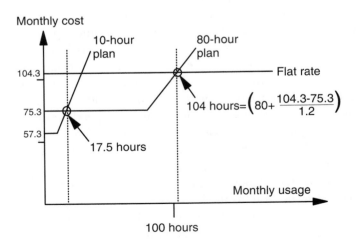

Figure 16.9 Three ISDN tariff plans compared.

the month). The flat rate tariff becomes economical if you envisage more than 104 hours usage per month or more than 4.7 hours a day.

Packet Switching and PSTN (Volume Versus Time)

A direct comparison is not possible between the volume charges associated with a packet-switched network and the per-minute charges of the PSTN. Neither can you make a break-even analysis to find out how many hours a day you would need to be using the service to make one economical over the other. One method of comparison is to work out the break-even number of *screens per minute* that makes the PSTN more cost effective than packet switching. This method is appropriate if the application is database access (the type of service to which packet switching is best suited).

The figures below are based on a screen with 50% fill (i.e., 960 characters or 15 segments [divide by 64] or 1 KB). We have also assumed that there would be one segment transmitted to the host for every screen that came back, making a total of 16 chargeable segments of data for every screen of data. The tariffs in Table 16.4 are reused here.

- PSTN cost per minute = $0.58;
- X.25 cost per minute = $0.10;
- X.25 cost per screen (i.e., 16 segments) = $0.032;
- Break even = ($0.58 − $0.10) / $0.032 = 15 screens per minute.

This means that if we expect to look up more than 15 screens per minute, then the PSTN will work out cheaper than packet switching.

This type of analysis can also be used to compare mobile packet data networks (RAM, Ardis, CDPD—Section 9.5.5) with circuit-switched data over cellular.

16.6 Tariff Trends

Influencing Factors

Influencing factors include:

- *Competition:* There are three categories of competition—value-added-services competition, basic-service competition, and infrastructure competition. Many countries only allow competition under the first heading at present. There is, however, a worldwide trend towards opening up all telecommunication services to competition.

- *Regulations:* These are rules imposed on the telephone companies. The idea is to promote fair play for all involved; however, vested interests can get in the way of this objective.

- *Online entertainment* (such as video on demand): Requires a large amount of bandwidth at the customer's premises (1.5 to 10 Mbps depending on the video quality required). If video-on-demand services are to be sold to residential consumers, the price of high-bandwidth services would have to drop to something close to $2 per hour to make them competitive with video cassette rental. This could have a dramatic impact on the price of bandwidth.

- *The Internet:* The flat-rate tariffs commonly charged for Internet access are already having an impact on telephone companies. Most dramatic is the increase in the volume of local calls, which are seldom a source of profit for the telephone company. Second, email and Internet telephony are in direct competition with telephone company services. Three effects are possible: a reduction in telephone company international charges, an increase in telephone company local call charges, or an increase in Internet charges (telephone companies in the U.S. have proposed that ISPs pay the same taxes that long-distance carriers must pay).

In most parts of the world, up until the early 1990s, long-distance and international call charges have been used to subsidize the cost of local calls. Increased competition has resulted in an increase in the cost of local calls and dramatic reductions in the cost of international calls. These price adjustments are often referred to as *tariff rebalancing.* Rebalancing is likely to continue in most countries for a few years because the markets are not fully open to competition and in many cases the price of local calls is still low, being a compromise between the price the telephone company would like to charge and what the residential consumer considers to be a fair price (bearing in mind that these consumers have been accustomed to low local-call charges for a long time).

Figure 16.10 shows how international call charges for various European operators have both fallen and converged over the last 12 years. The chart shows the range of international tariffs in the EU. One of the major influencing factors here has been the development of competition in the form of callback or routing of international calls over another network.

Figure 16.11, however, shows how the overall cost of telephone service in the same countries has neither declined nor converged to the same extent. This is explained in part by the lack of competition in national voice telephony in Europe. A further factor is the increases in local-call charges over the period.

1996 ECU

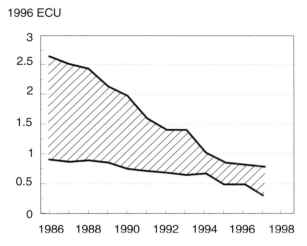

Figure 16.10 Price range for international calls—average cost per call minute for a six-line business customer. *Source:* [5] with permission.

1996 ECU

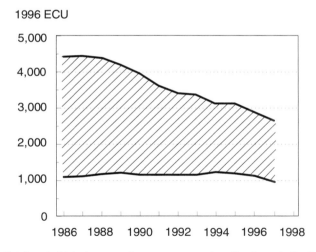

Figure 16.11 Total cost of telephone service—average cost per line for a six-line business customer. *Source:* [5] with permission.

Postalized Rates

There is a trend today towards charges becoming independent of distance—particularly for distances greater than about 50 km. This trend is seen across all services. For example, in many countries the cost of long-distance calls is dropping while the cost of local calls increases. Many data-communication services are now tariffed at rates that are independent of distance within national boundaries, and on the Internet this distance independence is global. This trend is partly due to the cost of transmission capacity becoming more and more distance independent, particularly in the case of optic-fiber and satellite links.

16.7 Sources of Tariff Information

Tariffs can always be obtained directly from the network operator. There are, however, a number of organizations that specialize in telecommunications tariffs. They are a convenient source of information when making tariff comparisons between operators and can provide software and tariff databases to ease the chore. Some of these organizations are listed below.

- Tarifica: London-based telecommunications tariff monitoring division of Phillips. Details on over 100 carriers worldwide (http://www.tarifica.com).
- Eurodata: a consultancy owned by European telecommunications operators (http://www.eurodata.co.uk).
- Ovum: trend reports with some tariff information (http://www.ovum.com).
- Valucom: service rates and rating elements for telecommunications carriers in over 50 countries worldwide (http://www.valucom.com).

References

[1]   Finnie, Graham, "Telco pricing falling into mixed basket—Complex tariff and feature packages are luring-and confusing-users," *Communications Week International,* May 22, 1995.

[2]   Tarifica: Telecommunications Tariffs and Consultancy, London.

[3]   Heywood, Peter, and Elke Gronert, "High Speed Data Services," *Data Communications International,* March 1996.

[4]   Greenfield, David, "Euro ATM—Get Ready for the Rollouts," *Data Communications International,* February 1997.

[5]   Louth, Graham, Federico Ciccone, and Lucy Landless (eds.), *Cutting the Cost—The Falling price of Telecom in Europe,* Cambridge, England: Analysys Publications, 1997.

Glossary

This glossary includes terms that appear in more than one chapter of the book. The definitions are not strictly precise; rather, they are designed to give newcomers a mental picture of common communications terms. More details can be found via the index.

Analog A method used to represent information such as speech. For example, in analog telephony, an electrical signal mimics the sound waves.

Asynchronous transfer mode (ATM) A networking technology supporting user bandwidths up to 155 Mbps (1997).

Automatic number identification (ANI) The same concept as caller ID. ANI is available to the owners of "800" and "900" numbers and emergency services ("911") in the United States. Unlike the newer caller ID service, the caller does not have the option to block ANI.

B-channel A channel for user information (voice or data) on an ISDN line.

Bandwidth The information-carrying capacity of a communication service or link.

Basic rate access (BRA) An ISDN line with two user channels. It is roughly equivalent to two PSTN lines; called BRI in the United States.

Basic rate interface (BRI) An ISDN line with two user channels; called BRA in Europe.

Bridge   A device for interconnecting local area networks.

Bus   A network topology where all the nodes are attached along a length of cable.

Call Center   An organization of people and technology whose primary function is to make and/or receive telephone calls (e.g., relating to sales or technical support).

Caller ID   The transmission of the caller's identity (telephone number) to the called party. See also ANI.

Calling line ID (CLID)   See Caller ID.

CCITT   An international standards body replaced by the ITU-T on March 1, 1993.

Central office   A public telephone switch; called an *exchange* in Europe.

Centrex   A service provided by a public switch that gives PBX facilities to the user.

Channel   A portion of the capacity of a communication link (e.g., thousands of speech *channels* can be carried over a single optic-fiber link).

Channel service unit/data service unit (CSU/DSU)   A device at the user end of a private line used to convert signals between the format used on the line and that used in the user equipment; called NTU in Europe.

Client server   Describes the situation where computer applications are split up between a server and the end user's computer (the client computer).

Closed user group (CUG)   A method of restricting access between users on a network. The basic idea being that members of a group can only establish communications with other members.

Computer telephony integration (CTI)   The linking of a phone system with a computer system so that the computer can control some of the telephony functions and the phone system can trigger events in the computer system.

D-channel   The signaling channel on an ISDN line.

Digital   A mode of representing information as a series of 1s and 0s so that it can be stored and manipulated with computers. Digital communication is the process of transmitting and switching this type of information.

Direct dial in (DDI)   See Direct inward dialing.

Direct inward dialing (DID)   A service that allows a caller from the public network to dial directly to a station (extension telephone) on a private telephone system (PBX).

Dual-tone multifrequency (DTMF)   The tone-signaling system used on many phones.

E1   A circuit with a bandwidth of 2 Mbps often divided into 30 (or sometimes 31) user channels of 64 Kbps each. E1 is used in Europe; see T1 for the U.S. equivalent.

Electronic data interchange (EDI)   A messaging system (based on email), where the messages are structured according to an agreed format so that they can be read by a machine. EDI facilitates electronic commerce.

Encryption   Manipulating (scrambling) information so that it becomes unintelligible to eavesdroppers but can be decrypted by a legitimate recipient.

Ethernet   A very popular type of local area network (LAN).

European Telecommunications Standards Institute (ETSI)   European standards body.

Exchange   A telephone switch; called a *central office* in the United States.

Fiber distributed data interchange (FDDI)   A type of local area network based on optic fiber. It operates at 100 Mbps and supports long (40-km) internodal links.

File transfer   Sending a computer file over a communication link.

Frame relay   A data networking technology supporting user bandwidths up to about 2 Mbps at present.

Global System for Mobile Communication (GSM)   A digital mobile phone system developed in Europe but used worldwide.

Information and communication technologies (ICT)   A general term covering all aspects of handling information electronically. The integration of information technology (IT) and communications technology.

Integrated services digital network (ISDN)   An enhancement of the telephone network that allows end-to-end digital communications at 64 Kbps and above.

Interactive voice response (IVR)   A system of "voice menus" on a phone system that can direct the caller to a specific destination or can give information to the caller. The caller selects options from the menus by dialing digits on their phone.

Interexchange carrier (IXC)   A U.S. term referring to a telephone company that provides long-distance service.

International Telecommunications Union—Telecommunications Standardization Sector (ITU-T)   An international standards body.

Internet   A worldwide computer network that is very widely available.

Internet Engineering Task Force (IETF)   A standards body that defines standards used by the Internet community.

Internet protocol (IP)   A communication protocol used on the Internet and on many private networks.

Intranet   A private network built using Internet technology.

Kilobits per second (Kbps)   One thousand bits per second; a measure of the information-carrying capacity of a communication link or the capacity requirements of a communication application.

Leased line   A private link between two sites; called a *private line* in the United States.

Local area network (LAN)   A computer network with limited geographical reach (e.g., usually within one building or campus).

Local exchange carrier (LEC)   A U.S. term referring to a telephone company that provides local service.

Local loop   The link between a services provider's nearest premises and your site.

Megabits per second (Mbps)   One million bits per second; a measure of the information-carrying capacity of a communication link or the capacity requirements of a communication application.

Modem   A device for sending digital signals on a communication link—commonly used to connect a PC to the telephone network.

Multiplexing   Combining a number of communication channels into an *aggregate* signal.

Network terminating unit (NTU)   A device at the user end of a private line (see CSU/DSU).

Node   A switch or router in any network; any computer or router on a computer network.

Open systems interconnect (OSI)   A framework for designing communication protocols established by the International Standards Organization (ISO).

Packet switching   A technique whereby information is routed through a network in small blocks called packets. It is very efficient for bursty (sporadic) traffic.

Permanent virtual circuit (PVC)   A permanent path through a data network.

Plain old telephone service (POTS)   The ordinary telephone service.

Primary-rate access (PRA)   An ISDN line with 23 (U.S.) or 30 (Europe) user channels or the equivalent of 23 or 30 PSTN lines. Called primary-rate interface (PRI) in the United States.

Private branch exchange (PBX)   A private telephone system typically used on a business premises. Sometimes called an electronic switchboard.

Private line   A private communication link between two sites. Also called a *leased line.*

Protocol   A set of rules governing the transfer of information in a network. Protocols may be implemented in hardware or software in the network nodes or end-user equipment.

Public switched telephone network (PSTN)   The ordinary telephone network.

Pulse code modulation (PCM)   A widely used method of converting an (analog) audio signal into a digital signal. In telephony the digital signal produced by PCM has a bit rate of 64 Kbps.

Regional Bell operating company (RBOC)   Any one of seven telephone service companies formed when deregulation forced the breakup of the Bell telephone monopoly in the United States.

Regulation   The process whereby national or supranational governments make rules about the provision and use of services. These rules generally promote fair competition, universal service, harmonization of services, and safety.

Request for information (RFI)   A formal request to a supplier for information.

Request for proposal (RFP)   A formal request to a supplier to propose (and price) a solution to a problem. Also called an invitation to tender.

Request for quotation (RFQ)   A formal request to a supplier to quote for goods or services.

Router   A device for interconnecting local area networks.

S-bus   The customer premises wiring used for basic rate ISDN.

Server   A computer that provides services to one or more end users' computers. Examples include web servers, file servers, and fax servers.

Signaling   The sending of control messages from one device to another (e.g., the sending of a dialed number from a telephone to the central office or exchange).

Service-level agreement (SLA) An agreement between a service provider and a client defining the service and how deviations from the service will be handled.

Switched multimegabit digital service (SMDS) A data networking technology supporting user bandwidths up to 25 Mbps at present (1997).

Synchronous digital hierarchy (SDH) A transmission and multiplexing system designed for high-capacity communication links (mainly optic fiber but also radio and copper). SDH is an international standard, which evolved from SONET.

Synchronous optical network (SONET) A transmission and multiplexing system developed in the United States and designed for high-capacity communication links (normally optic fiber or radio).

T1 A circuit with a bandwidth of 1.5 Mbps divided into 24 channels of 64 or 56 Kbps each.

Telco A telephone company.

Terminal adapter (TA) A device used in ISDN for connecting equipment (e.g., PCs) to the network.

Time-division multiplexing (TDM) A multiplexing technique used in digital systems where individual channels share the aggregate channel by time sharing.

Token ring A type of local area network.

Topology The layout of a network.

Traffic Information flow; this term is often used when referring to the *volume* of information flow.

Transmission control protocol (TCP) A communication protocol used on the Internet and on many private networks.

Transmission control protocol/Internet protocol (TCP/IP) A set of protocols used on the Internet and on many private networks.

Trunk A link between two switches.

UNIX   A computer operating system commonly found on Internet servers.

Very small aperture terminal (VSAT)   A small satellite dish (1m to 2m in diameter).

Virtual private network (VPN)   A networking solution based on public communication services that mimics some of the characteristics of private networks.

Voice mail   A phone messaging system that gives "answering machine" facilities to users.

Wide area network (WAN)   A network that goes beyond the confines of a premises or campus.

Workstation   An end user's computer.

World Wide Web (WWW)   The Internet as seen through a browser such as Microsoft Internet Explorer or Netscape Navigator.

X.25   A packet-switching technology used in low-speed (typically up to 64 Kbps) data networks.

About the Authors

Richard Downey joined the Irish Department of Post and Telegraphs* in 1981 as an executive engineer in the switching department, where he worked on traffic engineering, network synchronization, and exchange procurement. In 1988 he transferred to the technical training division, where he designed and delivered a range of courses on PCs and LANs in conjunction with managing the division's LAN and information systems. This was followed by the development and delivery of technical induction training for new telecommunication engineers in Telecom Eireann, data communication training for internal and external customers, and the design of documentation standards for the training division. He has a bachelor's degree in electrical engineering from the University of Dublin. Mr. Downey can be reached at rdowney@telecom.ie.

Seán Boland joined the Irish Department of Post and Telegraphs* in 1977 as a trainee technician. After he qualified, he worked as a technician on switching systems and customer circuit testing. During this time, he continued his studies privately and received the full technological certificate from the City and Guilds of London Institute. He was subsequently awarded a scholarship by Telecom Eireann and, in 1988, obtained an honors degree in electronic engineering from the University of Dublin.

His initial appointment was as executive engineer in Telecom Eireann's technical training division. During this period, he designed and delivered a wide range of courses dealing with communications technology. He also developed a close working relationship with Telecom Eireann's consultancy wing. This led to the development of many customized courses for external clients from many backgrounds and industries. It also lead to the development of the "Communications Technology" seminar, which was the inspiration for this book.

Seán subsequently moved to the Network and Group Technology Directorate within Telecom Eireann. His current responsibilities include evaluation of new technologies, network development, and technical support. Seán also has wide experience as a lecturer and examiner in technical subjects. He has been an open learning tutor with the University of Limerick, and he currently lectures in the Dublin Institute of Technology and the University of Dublin. His email address is sboland@telecom.ie.

Philip Walsh currently works with Ericsson Expertise in Dun Laoghaire, Ireland. He joined the Irish Department of Post and Telegraphs* as a trainee technician in 1980. In 1983 he was awarded a scholarship by the department to study electronic engineering in the Dublin Institute of Technology. On graduating in 1987 with an honors degree, he was appointed as an executive engineer in the technical training division in Telecom Eireann. Between 1987 and 1997, he was responsible for technology-based training, course design, PC applications training, external plant training, and switching training at various times within Telecom Eireann. In May 1997, he joined Ericsson Systems Expertise as an operations and maintenance consultant specializing in intelligent networks. Mr. Walsh can be reached at eei.eeipwalh@mesmtpse.ericsson.se.

---

* The Irish Department of Post and Telegraphs was split up in 1984. All three authors transferred into Telecom Eireann (the telecommunication company formed during the split) in 1984.

Index

The Artech House Telecommunications Library

Vinton G. Cerf, Series Editor

UNIX Internetworking, Second Edition, Uday O. Pabrai

Videoconferencing and Videotelephony: Technology and Standards, Richard Schaphorst

Voice Recognition, Richard L. Klevans and Robert D. Rodman

Wireless Access and the Local Telephone Network, George Calhoun

Wireless Communications in Developing Countries: Cellular and Satellite Systems, Rachael E. Schwartz

Wireless Communications for Intelligent Transportation Systems, Scott D. Elliot and Daniel J. Dailey

Wireless Data Networking, Nathan J. Muller

Wireless LAN Systems, A. Santamaría and F. J. López-Hernández

Wireless: The Revolution in Personal Telecommunications, Ira Brodsky

Writing Disaster Recovery Plans for Telecommunications Networks and LANs, Leo A. Wrobel

X Window System User's Guide, Uday O. Pabrai

For further information on these and other Artech House titles, including previously considered out-of-print books now available through our In-Print-Forever™ (IPF™) program, contact:

Artech House
685 Canton Street
Norwood, MA 02062
781-769-9750
Fax: 781-769-6334
Telex: 951-659
email: artech@artech-house.com

Artech House
Portland House, Stag Place
London SW1E 5XA England
+44 (0) 171-973-8077
Fax: +44 (0) 171-630-0166
Telex: 951-659
email: artech-uk@artech-house.com

Find us on the World Wide Web at:
www.artech-house.com